Information Systems Engineering and Management

Volume 82

The book series "Information Systems Engineering and Management" (ISEM) publishes innovative and original works in the various areas of planning, development, implementation, and management of information systems and technologies by enterprises, citizens, and society for the improvement of the socio-economic environment.

The series is multidisciplinary, focusing on technological, organizational, and social domains of information systems engineering and management. Manuscripts published in this book series focus on relevant problems and research in the planning, analysis, design, implementation, exploration, and management of all types of information systems and technologies. The series contains monographs, lecture notes, edited volumes, pedagogical and technical books as well as proceedings volumes.

Some topics/keywords to be considered in the ISEM book series are, but not limited to: Information Systems Planning; Information Systems Development; Exploration of Information Systems; Management of Information Systems; Blockchain Technology; Cloud Computing; Artificial Intelligence (AI) and Machine Learning; Big Data Analytics; Multimedia Systems; Computer Networks, Mobility and Pervasive Systems; IT Security, Ethics and Privacy; Cybersecurity; Digital Platforms and Services; Requirements Engineering; Software Engineering; Process and Knowledge Engineering; Security and Privacy Engineering, Autonomous Robotics; Human-Computer Interaction; Marketing and Information; Tourism and Information; Finance and Value; Decisions and Risk; Innovation and Projects; Strategy and People.

Indexed by Google Scholar. All books published in the series are submitted for consideration in the Web of Science.

For book or proceedings proposals please contact Alvaro Rocha (amrrocha@gmail.com).

Gisela Haro Esquivel · Pablo Ayala Hernández

Bridging the Gap: Comprehensive Project Design for Soft Skills Development and Labor Market Integration

Springer

Gisela Haro Esquivel
Juárez, Chihuahua, Mexico

Pablo Ayala Hernández
Juárez, Chihuahua, Mexico

ISSN 3004-958X ISSN 3004-9598 (electronic)
Information Systems Engineering and Management
ISBN 978-3-032-19025-3 ISBN 978-3-032-19026-0 (eBook)
https://doi.org/10.1007/978-3-032-19026-0

This Springer imprint is published by the registered company Springer Nature Switzerland AG
The registered company address is: Gewerbestrasse 11, 6330 Cham, Switzerland

We dedicate this work with gratitude to our children, students, and friends, from whom we continuously learn and to whom we teach. Their support and example have fostered our academic growth and inspired this book.

Contents

Abbreviations, Acronyms and Initialisms

CAN	National Accreditation Commission
CON reg	Document Registered with the United States Navy
CONARC	National Council for the Registration of Professional Certification of Architects
Donamis	Power, Strength or Skill, Services and Projects Platform
empleosmaquila	Maquila Jobs, Job Search Platform
ENOE	National Survey of Occupation and Employment
Et al.	And Others
HB	Soft Skills
HD	Hard Skills
IL	Labor Insertion
IELMEC	Electromechanical Engineering
Indeed	Ingresa, Job Search Platform
INEGI	National Institute of Statistics and Geography
ITCJ	Technological Institute of Cd. Juárez
OCC	Online Career Center, Job Search Platform
PHB	Soft Skills Question
PIL	Labor Insertion Question
Redalyc	Network of Scientific Journals of Latin America, the Caribbean, Spain, and Portugal
S.A.	Corporation (Inc.)
Scielo	Scientific Electronic Library Online (SciELO)
SEP	Secretariat of Public Education
SSRN	Social Science Research Network (SSRN), Information Journal Platform
STPS	Secretariat of Labor and Social Welfare
TecNM	National Technological Institute of Mexico
UPM	Polytechnic University of Madrid
UTN	National Technological University

List of Figures

List of Tables

Part I
Soft Skills and Their Relationship with the Labor Market Integration

This part of the study explores the relationship between soft skills (SS) and labor market integration (LMI) among students of the Electromechanical Engineering program (IELMEC) at the Technological Institute of Ciudad Juárez (ITCJ). Findings reveal a mismatch between professional training and employment opportunities, as a large proportion of graduates work outside their field of specialization. National statistics and labor market data confirm that over half of professionals face this situation, often due to limited opportunities or insufficient career planning.

Employers increasingly demand both technical skills (TS) and soft skills (SS). While TS are formally acquired through academic training, SS are often informally developed and play a decisive role in employability. However, IELMEC students do not consistently develop these skills within their formal education, which contrasts with the graduate profile defined by the National Technological System of Mexico (TecNM).

The research is structured into five chapters: identification of the problem and objectives, theoretical framework, methodology (including questionnaires and interviews), data analysis of student responses, and conclusions. The study emphasizes the importance of strengthening soft skills development within engineering education to enhance professional integration and better align graduates' competencies with labor market demands.The research is structured into five chapters: identification of the problem and objectives, theoretical framework, methodology (including questionnaires and interviews), data analysis of student responses, and conclusions. The study emphasizes the importance of strengthening soft skills development within engineering education to enhance professional integration and better align graduates' competencies with labor market demands.

Chapter 1
Introduction to the Problem

This study was conducted with the purpose of examining the relationship between soft skills (SS) and labor market insertion (LMI) among Electromechanical Engineering (IELMEC) students at the Technological Institute of Ciudad Juárez (ITCJ). The analysis focuses on understanding how these concepts interact in a context where, in Mexico, a significant percentage of graduates are unable to secure employment in their field of specialization due to limited labor market opportunities.

1.1 Introduction

According to data from OCCMundial, the largest job portal in Mexico, in 2018, five out of ten professionals did not work in their area of study. Additionally, 88% of respondents recognized that their studies contributed to their professional development; however, 55% reported having changed their specialty due to lack of opportunities, and 34% due to insufficient planning of their professional future. According to the National Institute of Statistics and Geography [1], during the third quarter of 2019, Mexico had 15.09 million professionals, of whom 11.79 million were employed. Nevertheless, 5.42 million of these worked in areas unrelated to their academic training, representing 54% of employed professionals working outside their specialization.

Currently, employers demand that their future employees possess technical skills, also known as hard skills (HS), which are predominantly acquired through formal education in academic institutions. However, employers also seek soft skills (SS), which are not systematically taught in classrooms but develop more organically without explicit awareness in students. This imbalance is reflected in the graduate profiles of IELMEC students within the National Technological Institute of Mexico (TecNM) system, to which ITCJ belongs. Table 1.1 details the general objective of the program as well as the entry and exit profiles of its students, highlighting that over

G. H. Esquivel and P. A. Hernández, *Bridging the Gap: Comprehensive Project Design for Soft Skills Development and Labor Market Integration*, Information Systems Engineering and Management 82, https://doi.org/10.1007/978-3-032-19026-0_1

70% of the evaluated competencies correspond to hard skills, while the development of soft skills receives considerably less attention during academic training.

What is the relevance of these topics in the Electromechanical Engineering (IELMEC) program?

This question constitutes precisely the rationale of the present research, whose objective is to analyze the relationship between soft skills (SS) and labor market insertion (LMI). Faculty members of the program have observed that, although students often encounter difficulties in securing employment that allows them to fully develop professionally, companies recognize that graduates possess excellent mastery of the technical skills aspects related to their specialization. However, employers also emphasize that graduates demonstrate certain limitations in the social domain.

Companies seek engineers who not only possess technical skills but also soft skills, since the latter are fundamental in professional life. Specifically, soft skills enable engineers to communicate clearly, directly, and empathetically with their subordinates, thereby improving effectiveness in daily work. An example of this corporate demand can be observed in the way the company AUTOMOTIVE advertises a vacancy for an APQP Engineer within its organization.

Job description: Automotive is a global leader in solutions configuration and manufacturing components that enable connectivity in automotive systems. With presence in more than 40 countries, the company allocates its resources daily to one purpose: to support clients in achieving their success. Position offered: APQP

Table 1.1 Electromechanical engineering [2]

General objective	To train excellent professionals in Electromechanical Engineering with an entrepreneurial attitude, leadership skills, and the ability to analyze, diagnose, design, select, install, manage, maintain, and innovate electromechanical systems efficiently, safely, and economically, considering national and international standards to promote sustainable development with full ethical, humanistic, and social awareness
Admission profile	Graduates possess leadership abilities and an entrepreneurial vision for problem-solving and prevention, developing innovations in productive and service sectors. They are capable of managing projects, human, material, and financial resources efficiently. They receive a comprehensive education, are socially aware, responsible, participative, and possess effective communication skills to interact with interdisciplinary teams and apply acquired knowledge in analysis, diagnosis, evaluation, and problem-solving within organizations. They utilize advanced technologies with a vision of competitiveness and quality.
Graduate profile	Graduates can design, manage, and evaluate engineering projects related to electromechanical systems and devices, proposing solutions using cutting-edge technologies within the framework of sustainable development. They can design and implement electromechanical systems and devices, optimizing energy use in productive and service sectors according to national standards and agreements. They develop an entrepreneurial attitude by creating and incubating companies, innovating products and services in the electromechanical sector. They apply computational software tools for the design, simulation, and operation of electromechanical systems according to industrial demand. Graduates can interpret, understand, and communicate ideas, texts, and manuals in a second language.

Engineer, Juárez Plant. An engineer in APQP (Advanced Product Quality Planning) at Automotive is responsible for ensuring that new or modified products meet quality requirements and are launched efficiently and seamlessly. APQP is a structured methodology that guides the product development process in the automotive industry, and the engineer in this area holds several key responsibilities.

Experience requirements: Completed professional degree; proficiency in Microsoft Office; file and document management; at least 80% proficiency in written and spoken English; knowledge and management of SAP and BEAMS systems; interpretation of harness and component drawings; proven experience leading internal and external meetings (BU); and one year of verifiable work experience.

Skills requirements: Leadership, organizational capacity, decision-making, teamwork, proactivity, effective communication, and punctuality.

In the job offer from Automotive, it becomes evident that companies value both technical competencies and soft skills as key requirements for successful entry into the labor market.

The Technological Institute of Ciudad Juárez (ITCJ), part of the National Technological Institute of Mexico (TecNM) system, offers several degree programs in Ciudad Juárez, including Business Administration, Accounting, Industrial Engineering, Business Management Engineering, Logistics Engineering, Electrical Engineering, Electronic Engineering, Computer Systems Engineering, Mechatronics Engineering, Mechanical Engineering, and Electromechanical Engineering. The latter constitutes the primary focus of this research, which seeks to analyze why some students integrate easily into the local labor market, while others, despite having outstanding grades, fail to secure employment in their field of specialization.

To determine the relationship between soft skills (SS) and labor market insertion (LMI), an analysis was conducted with the participation of seventh- and eighth-semester students, who are about to enter the labor market, as well as program graduates currently employed in the local industry. The sample analyzed included both those who participated in student projects and those who did not, in order to assess whether the development of soft skills has been a determining factor in facilitating their insertion into the labor market.

It is often argued that participation in student projects, student chapters, or career clubs within the classroom contributes to the development of soft skills and strengthens students' social abilities. However, to what extent is this assertion valid? Over time, faculty members of the Electromechanical Engineering program (IELMEC) at ITCJ have observed that students involved in classroom projects achieve a smoother integration into the local industry. This research aims to clarify whether soft skills are indeed a key factor determining the success of graduates in their entry into the labor field.

The teaching-learning process in engineering is constantly seeking mechanisms to adapt to the evolution of technological and social changes. Therefore, analyzing and adopting new methodologies is not only a common practice but also essential to train engineers who meet the expectations of the labor market and who can interact effectively within a constantly evolving world.

Currently, the global population operates in a highly interconnected environment, where the success of professionals depends on acquiring new tools, skills, and knowledge. This process begins in academic institutions, is reinforced through early work experiences, and is consolidated through continuing education.

Among these new methodologies, active or project-based methodologies stand out, whose purpose is to teach students how to "be," "belong," and "do," addressing aspects related to procedures, attitudes, and learning concepts. These strategies promote the development of skills that enable students to evaluate the complexity of problems, critically understand information, and assess their progress in knowledge acquisition. Likewise, they foster individual commitment, positioning students as the main actors in the teaching-learning process.

1.2 Justification

The execution of this research is crucial, as the analysis will allow the establishment of the relationship between soft skills (SS) as a useful tool for labor market insertion (LMI) of graduates from the Electromechanical Engineering (IELMEC) program at ITCJ within the local industry.

Soft skills play a fundamental role for IELMEC students who, although they develop technical skills (HS) on a daily basis, often neglect the development of soft skills. The objective of this research is to encourage students to actively promote the advancement and growth of their soft skills so that, when seeking employment, these abilities may facilitate or enhance their insertion into the local labor market.

When IELMEC students and graduates with well-developed soft skills who enter the city's labor market often occupy positions requiring constant interaction with others, such as area coordinators, supervisors, or managers. In these roles, soft skills are essential for fostering a harmonious and collaborative work environment and ensuring effective organizational management. This not only improves personal and professional performance but also contributes to a more efficient and cooperative workplace for colleagues and collaborators.

1.3 Problem Statement

The National Technological Institute of Mexico (TecNM) comprises 254 institutes across the national territory, making this system the largest in Latin America. In the state of Chihuahua, there are nine institutes, and in Ciudad Juárez, there are two: the Technological Institute of Ciudad Juárez (ITCJ) and the Technological Institute of Ciudad Juárez Campus II (ITCJ Campus II). These institutions offer a variety of academic programs ranging from undergraduate to doctoral levels. The undergraduate program in Electromechanical Engineering (IELMEC) is offered at ITCJ, and this research focuses on students who, during 2023, were enrolled in the seventh and eighth semesters of this program, considering a sample of ten students as well as five graduates from different cohorts who are currently working in the local industry.

Given that IELMEC is a program primarily grounded in the development of technical knowledge, often overlooking the social dimension while emphasizing the advancement of hard skills (HS), it is important to analyze whether the soft skills (SS) developed by students actually benefit them in entering the labor market within their area of specialization. The purpose of this study was to examine the relationship between the soft skills developed by students in the Electromechanical Engineering program at the Technological Institute of Ciudad Juárez (ITCJ) and their subsequent labor market insertion. While it is often argued that participation in projects within academic courses or student chapters has enhanced students' daily performance, the central question arises: **to what extent is this truly the case?**

1.4 Research Question

What is the relationship between soft skills, considered as useful tools, and the process of labor market insertion of Electromechanical Engineering students at the Technological Institute of Ciudad Juárez (ITCJ)?

1.5 Assumption

It is assumed that the development of soft skills (SS) among students constitutes a valuable tool that positively influences employers' hiring decisions. This, in turn, facilitates the labor market insertion of these students into the local industry by making them more attractive as candidates who demonstrate essential interpersonal and social competencies required in the workplace.

1.6 Objectives

The following objectives are established to provide a clear direction for the study and to define the scope of the analysis. They aim to highlight the relevance of soft skills in the professional training of university students and their impact on labor market insertion, with particular attention to the case of Electromechanical Engineering.

1.6.1 General Objective

To identify the relationship between soft skills (SS), considered as useful tools, and the labor market insertion (LMI) of university students.

1.6.2 Specific Objectives

1. To determine, through the validated questionnaire on SS, which soft skills are most developed by Electromechanical Engineering students.
2. To identify the actual graduation characteristics and which soft skills facilitated their entry into the professional and labor market.

1.7 State of the Art: Soft Skills (SS) and Labor Insertion (LI)

At present, education is grounded in the comprehension of learning and the development of competencies. The main actors in this Teaching–Learning process are teachers and students, who employ diverse resources to keep learners engaged in their learning journey. The ultimate goal is to achieve the specific competence established in each subject, while, as a collateral and synergistic effect, students develop and acquire professional skills and abilities that will serve them in their professional lives [3], particularly for the graduates of the Technological Institute of Ciudad Juárez (ITCJ).

The resolution of specific technical problems provides the opportunity for students to autonomously develop the learning of professional competencies, offering them a clearer perspective of their potential actions within the Electromechanical Engineering (IELMEC) program.

Regarding faculty members, they have access to a wide range of methodologies, among which the active methodology—also referred to as project-based learning—stands out. This approach assigns students the central role in the teaching–learning process. It challenges them to learn and assumes that knowledge acquisition is achieved by the students themselves, while the teacher's role is limited to guiding and orienting learners toward the achievement of the intended objectives. This implies a shift in the student's role, moving from being a passive recipient of knowledge to an active participant required to communicate, analyze, decide, verify, and critically examine their own learning. Evaluation, accordingly, undergoes modifications, being based on the student's process throughout the course.

Effective and immediate communication between teacher and student is essential. Therefore, the instructor must be willing to accompany the learner during the development of their learning through projects.

This type of active model enables students to construct their own knowledge and provides them with the necessary tools to understand the strategies required to achieve their goals. Such processes enhance decision-making efficacy, improve reflective capacity regarding the learning process, and foster the creation of knowledge through experiential learning. Moreover, students implicitly develop soft skills such as oral and written communication, teamwork, and interpersonal communication.

In alignment with the writing and intended audience of this study, it is necessary to establish what constitutes soft skills and labor insertion, so that the data collected and the subsequent analysis yield truly meaningful outcomes.

1.7.1 Soft Skills (SS) and Hard Skills (HS)

Soft skills emerge from a combination of experience, social learning, and professional development. Nowadays, they are widely recognized in corporate recruitment and selection processes, given that they are acquired through both professional and personal life experiences of applicants and are considered essential for determining success in the workplace.

The origin of soft skills can be traced back to the 1950s, when technological training processes demanded considerable resources from the United States Army. A decade later, the U.S. Army developed a training systems engineering program, formalized in the official document *CONReg 350-100-1*, which specified the type of training and instruction required for military personnel to develop the minimum knowledge and skills necessary for optimal task performance.

The term *soft skills* was formally introduced for the first time in 1972 in the U.S.Army manual and presented at the "CONARC Soft Skills Training" conference by Dr. Paul G. Whitmore, as part of his thesis entitled *What are Soft Skills?* In this work, Whitmore described soft skills as interpersonal and organizational competencies involving roles and responsibilities such as inspection, personnel office management, planning documentation, efficiency reports, and the design of bridge and barracks structures. Soft skills enable individuals to establish, understand, expand, and strengthen interpersonal relationships. Among the most relevant are:

1. Time management—effective organization to accomplish assigned tasks, thereby increasing work performance.
2. Critical thinking—directly influencing problem-solving and decision-making.
3. Change management—the ability to adapt to new organizations and their policies.
4. Stress management—the ability to handle unexpected or challenging situations effectively.
5. Communication skills—the capacity to convey ideas and thoughts (oral and written) in an active and assertive manner, including language-related competencies.
6. Leadership—the ability to influence, motivate, and organize individuals to achieve objectives set by either the leader or the organization.

7. Resilience—the capacity to adapt to difficult situations, remain steadfast, and grow in the face of adversity.
8. Cognitive or affective empathy—the ability to understand others' thoughts and feelings.

Thus, soft skills are developed through lived experiences, professional contexts, and interpersonal interactions, which foster individual growth and distinguish individuals in social and professional environments.

Gallardo [4] notes that soft skills result from the integration of diverse abilities such as social skills, communication, personal development, leadership, creativity, teamwork, emotional intelligence, and critical thinking, all of which are directed toward facilitating relationships and ensuring effective project execution (*Gestion, Management & Employ*).

Similarly, Coll Morales [5] defines soft skills as those that enable individuals or professionals to integrate seamlessly into their environment. These include teamwork, effective communication, and the ability to interact with diverse individuals. The main soft skills relevant to this study include:

1. Effective communication.
2. Punctuality.
3. Emotional intelligence.
4. Cooperation.
5. Adaptability.
6. Conflict resolution.
7. Stress management.
8. Time management.
9. Proficient oral and written communication.
10. Teamwork.
11. Critical thinking.
12. Sociability.
13. Creativity.
14. Organizational skills.

In contrast, hard skills refer to technical competencies, typically acquired through academic training in professional programs, which serve as the tools to perform specific tasks within one's field.

According to Coll Morales [5], hard skills are the professionalizing competencies individuals acquire throughout their academic and professional trajectories, enabling them to perform their work effectively. Examples include:

1. Writing proficiency.
2. Mathematical and computational skills.
3. Proficiency in office software.
4. Operation of computer systems.
5. Complex problem management.
6. Resource optimization and maximization.
7. Operations management.

8. Personnel management.
9. Effective handling of sensitive situations.
10. Development of large-scale projects.
11. Operation and management of machinery and tools.

1.7.2 Labor Insertion (LI)

Currently, engineering graduates are subject to greater scrutiny and are expected to possess a wider range of skills and competencies to secure the opportunity for meaningful employment and to succeed in their professional field. Modern employers seek engineers who demonstrate both soft skills (SS) and hard skills (HS).

1. Teamwork—This skill not only involves the willingness to collaborate with others but also encompasses the characteristics that make teamwork successful:
   (a) Establishing and communicating common goals—Professionals recognize the objectives, understand their importance, and effectively convey them to the team, defining goals and the pathway to achieve them.
   (b) Respecting and valuing the work of others—Professionals understand that no task is insignificant; individual contributions are valuable and are acknowledged, especially upon reaching objectives.
   (c) Exercising leadership—Professionals guide and motivate colleagues, emphasizing service and support for the team rather than exercising authority.
   (d) Communicating effectively—Coordinating efforts through clear and efficient communication, encouraging and recognizing team members for providing progress reports, identifying problems, and proposing solutions, thereby enriching the communication process.
2. Creativity and process innovation—Engineers must optimize processes and tasks to perform effectively in any area:
   (a) Optimizing resource utilization—Minimizing waste and losses.
   (b) Enhancing production efficiency—Maximizing the value of time.
   (c) Designing processes to optimize machinery and workflow.
   (d) Ensuring maximum profitability.
3. Self-directed learning—Engineers require the skill and willingness to continuously learn and grow both personally and professionally. Problem-solving necessitates a commitment to continuous improvement and the pursuit of excellence.
4. Adaptability to change—Engineers must be prepared to adjust and reform professional and personal paradigms as needed. The only constant in contemporary professional environments is change; thus, the ability to adapt and pursue greater and better opportunities is imperative.

5. Collaboration in multidisciplinary teams—Historically, engineers have been characterized as individualistic professionals focused on hard skills, technical knowledge, and creativity. However, in practice, engineers frequently work in multidisciplinary teams to serve diverse clients. The ability to collaborate effectively with a team is a critical factor for employability.

As observed, these skills are not solely technical; they develop over time through experience and life exposure. A newly graduated student who has focused exclusively on academic studies may not have acquired these skills without practical work experience. Employers recognize this gap and evaluate candidates' curricula and interview performance to determine whether they possess the minimum soft skills required to integrate into a team and acquire the competencies necessary to establish themselves positively within the organization.

References

1. National Institute of Statistics and Geography (INEGI).: National Survey of Occupation and Employment (ENOE). Statistics of Professionals in Mexico (2020). https://www.inegi.org.mx/contenidos/saladeprensa/aproposito/2020/EAP_EconFin.pdf
2. TecNM-Academy of Mechatronics.: Engineering in Mechatronics (2020). http://cdjuarez.tecnm.mx/programas_academicos/
3. Valle, A., González, R., Lino, C., Cuevas González, M., Fernández Suárez, A.P.: Learning strategies: basic characteristics and their relevance in the school context **6**, 53–68 (1998). https://www.redalyc.org/pdf/175/17514484006.pdf
4. Gargallo, A.: Active methodologies in strategic business management: student involvement in university strategic analysis. RIE **5**(48), 1–8 (2009)
5. Coll Morales, F.: Soft skills. Economipedia.com (2021). https://economipedia.com/definiciones/habilidades-blandas.html

Chapter 2
Theoretical Framework

The relationship between soft skills and labor market insertion constitutes the primary focus of this research. Within this framework, it is essential to distinguish between soft skills and hard skills, as both play a decisive role in shaping professional performance and employability. Hard skills refer to the technical knowledge and competencies acquired through formal education and specialized training, which are indispensable for practicing engineering in industrial contexts. In contrast, soft skills encompass interpersonal, cognitive, and attitudinal abilities—such as communication, teamwork, adaptability, and leadership—that complement technical expertise and are increasingly valued in professional environments.

For students of Electromechanical Engineering (IELMEC), the integration of these two dimensions is particularly relevant given the demands of the contemporary labor market, which requires not only technical proficiency but also the capacity to address complex challenges collaboratively and creatively. Consequently, the theoretical framework presented in this chapter provides the conceptual basis for analyzing how the development of both soft and hard skills influences the successful labor market insertion of IELMEC graduates, as well as their long-term professional growth and adaptability in dynamic industrial settings.

2.1 Soft Skills (SS)

Since not all individuals are familiar with the term soft skills (SS), it is necessary to clarify this concept and, above all, examine how these skills can assist or facilitate university students—specifically, Electromechanical Engineering (IELMEC) students at the Technological Institute of Ciudad Juárez (ITCJ)—in their process of entering the labor market, which may begin as early as the seventh semester or continue until the completion of their academic program.

G. H. Esquivel and P. A. Hernández, *Bridging the Gap: Comprehensive Project Design for Soft Skills Development and Labor Market Integration*, Information Systems Engineering and Management 82, https://doi.org/10.1007/978-3-032-19026-0_2

What are soft skills? Soft skills are characteristics, attitudes, behaviors, and personality patterns that enable individuals to socialize effectively and manage life situations with a certain degree of emotional control. These skills may be learned or innate. Learned skills are those developed throughout an individual's life, through experience or study, whereas innate skills are those with which an individual is born.

Thus, which are the natural skills that individuals are born with, and which are those that can be developed over the course of life? Often, people observe characteristics in others, including young children, and describe them with expressions such as:

1. What an "extroverted" child.
2. That girl "communicates" very well with adults and other children.
3. You are a very "sharing" child.
4. What a "creative" girl.
5. So "curious."
6. That boy is a "leader."
7. How "sensitive" he is to others.
8. Among others.

These skills are interpersonal social skills, and they pertain to the manner in which individuals interact with others, how they treat, collaborate, or act in situations that require dealing with one or more people. Nevertheless, many of the aforementioned skills can be acquired through experiences gained over the course of life, not necessarily within the academic environment.

Soft skills, also known as transversal competencies, can be innate in each individual, but they can also be developed and cultivated to their fullest potential. They provide an additional benefit in the workplace and encompass values, personal characteristics, and acquired knowledge. Although they are common across all job positions, they are highly valued because they can significantly impact organizational productivity.

For the purposes of this research, only learned soft skills will be considered, as these are accessible to any individual through life experiences or formal educational processes.

The origin of soft skills can be traced back to the 1950s, when technology training processes consumed considerable resources within the United States Army. A decade later, the national navy of the United States developed a training systems engineering program, officially documented in CON Reg 350-100-1, which specified the type of training and education necessary for naval personnel to develop the essential skills and knowledge for optimal task performance.

The term soft skills was formally documented for the first time in the United States Army Manual in 1972 and presented at the "CONARC Soft Skills Training" conference by Dr. Paul G. Whitmore, in his thesis titled "*What Are Soft Skills?*", where he noted that soft skills involve individuals and roles required to inspect, manage personnel offices, conduct studies, document maintenance plans, efficiency reports, or design bridge and barracks structures. Soft skills enable individuals to establish, understand, expand, and strengthen relationships with others. Some of the soft skills that facilitate this include:

1. Empathy: the individual's capacity to understand the life situation of others.
2. Time management: the proper administration of time to carry out the assigned task, which enhances work performance.
3. Critical thinking: this type of thinking directly impacts problem-solving and aids in decision-making.
4. Change management: the ability to adapt effectively to a new organization and its policies.
5. Stress management: the manner in which problems, unforeseen situations, or difficult circumstances are addressed.
6. Communication skills: the capacity to convey ideas and thoughts in all forms (written or oral) in an active and assertive manner. This category includes language skills.
7. Leadership: the ability to influence, motivate, and organize individuals in order to achieve objectives set by the leader or the organization.
8. Resilience: the capacity to adapt to difficult situations, remain steadfast, and grow in response to challenges.
9. Cognitive or affective empathy: understanding other people in terms of their thoughts and feelings.

Soft skills are therefore those abilities that an individual develops or cultivates through life experiences and the various situations they encounter. These skills help the individual to stand out personally among others. Some of the soft skills most frequently demanded by local companies or industries are presented in Fig. 2.1.

In Robbins' [1] article, "*Soft Skills and Success in the Workplace*", he mentions that soft skills propel individuals to develop more effectively in the professional sphere, projecting them toward success in life through socio-emotional interaction and positive influence. A significant portion of success and productive performance with collaborative groups depends on the soft skills of the team leader.

One of the primary authors who has explored the topics of soft skills and labor market integration or productive work is Goleman [2]. In his book *Social Intelligence*, he states that by developing interpersonal relationship capabilities, it is possible to establish important emotional bonds among individuals, an effect that replicates itself in what are known as virtuous circles. Therefore, according to Goleman, soft skills are the most effective way to engage in academic, professional, and personal projects and elevate them to a high-quality level while maintaining effective socio-affective relationships.

Consequently, it is essential to define some of the skills that make individuals competent in the labor market. According to Pita Yañez [3], in her book *How to Be Competent*, there are 18 competencies (Table 2.1) whose value concept has redefined professional profiles. As previously mentioned, these are the competencies that companies seek in their candidates. These competencies are not only based on knowledge and technical skills but also on the attitudes and personal values necessary for professional development.

Fig. 2.1 Most requested soft skills in the industry. Own elaboration

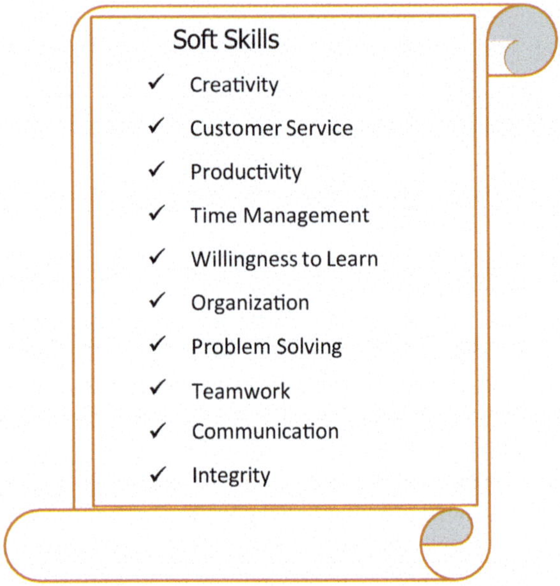

Soft skills can be learned, and the following points explain how soft skills can be acquired. Some of these points are established in the master's thesis in Human Resources and Organizational Management at the Huancayo Graduate School in 2021, regarding soft skills and job performance of administrative workers, by Chaca Oliveros and Contreras Tito:

1. Search: Having personal initiative in the acquisition, improvement, or transmission of a soft skill.
2. Information: This involves detailed description and clear communication of the foundations and principles associated with the skill in question, with the aim of facilitating understanding and assimilation by the learner. This process may include the explanation of key concepts, practical examples, and any relevant information that contributes to the effective acquisition of the skill.
3. Demonstration: Demonstrating skills through simulations involves recreating real-world situations, role-playing, or other environments that mimic scenarios relevant to the skill being taught. Instead of merely explaining concepts theoretically, learners are given the opportunity to experience and apply this knowledge in practical contexts. These simulations can take various forms, such as interactive scenarios, practical activities, or even games specifically designed to teach particular skills, leading to the next point.
4. Practice: The goal is to provide a space where students can put into practice what they have learned, face challenges similar to those they might encounter in real life and develop their skills and abilities effectively. This approach not only allows for a deeper understanding of the concepts but also promotes active learning by engaging students in the practical application of their knowledge.

Table 2.1 Competencies of a competent individual. Own elaboration

Skills	Characteristics
1. Self-awareness	• Confidence in one's own ability to carry out tasks, assume responsibilities, and face challenges • Influence over thoughts and behaviors that help achieve objectives, such as perseverance, effort, and commitment • Transversal competency or soft skill, one of the most important, as it refers to attitude or personality aspects
2. Analysis, synthesis and critical thinking	• Ability to decompose complex situations to understand their underlying principles or elements • Analysis enables drawing conclusions and making future projections • Synthesis allows the integration of parts into a coherent whole • Critical thinking involves analyzing and evaluating with the purpose of improvement This skill is essential in the business world, as it provides a unique and genuine value to organizations, distinguishing them from others
3. Organization and planning	• Define priorities in task execution by establishing action plans through the optimization of time, methods, and resources • Skill present across all areas of the organization, obviously applied at different levels depending on the position of responsibility within the area or department
4. Communications skills	• Exchange of messages for the transmission of information • Communication must occur within a context; context refers to the set of elements, factors, and circumstances that enable the act of communication • Communication is a strategic tool for organizations that require effective communication, teamwork, innovation and creativity, and leadership. It can be oral, written, or non-verbal • Communication consists of five steps: 1. Formulate the idea to be conveyed; 2. Translate the message into words, images, or gestures; 3. Transmit the message clearly; 4. Listen actively; 5. Receive feedback
5. Responsability and perseverance	• Responsibility is the capacity to recognize and accept the consequences of decisions and actions that have been taken • Perseverance is consistency, firmness, or tenacity in carrying out an action • The integration of individuals with this characteristic into organizations is a valuable asset, especially in times of uncertainty

(continued)

Table 2.1 (continued)

Skills	Characteristics
6. Decision-making	• The ability to identify and analyze problems in order to undertake actions that facilitate problem-solving, ensuring quality in the results • Decision-making that affects people is of utmost importance due to its level of complexity • This competency is one of the most critical for individuals aspiring to become directors or managers within a company
• Customer orientation	• The ability to respond to the current needs of the client, anticipate future needs, and establish a trust-based relationship between the company and the client • This competency enables individuals to identify, measure, and develop behaviors that result in positive performance in their role • Customer orientation is an attitude that should be adopted in all types of work
7. Teamwork and cooperation	• Teamwork consists of a group of individuals working together to achieve a common objective or mission, coordinating their efforts and complementing each other's skills • Teamwork enables success within organizations and can be considered a key factor for achievement • It requires a clear mission, definition of each member's roles and responsibilities, active participation from all members, and the establishment of a conflict resolution system • The characteristics of an "ideal team player" include: team spirit, collaboration, respectfulness, good character, responsibility, diligence, and a proactive attitude
8. Learning and adaptability	• The ability for continuous learning throughout life, as well as the capacity to adapt to new situations, being open to new experiences, recognizing the positive aspects of opportunities, and testing one's knowledge, skills, and abilities when facing new challenges • This is a soft skill, independent of professional profile, but it is crucial when seeking employment
9. Flexibility and change orientation	• The ability to understand different perspectives of a particular situation, in order to adapt and work effectively during moments of change

(continued)

Table 2.1 (continued)

Skills	Characteristics
	• Adaptability to change is a trait inherent to humans; however, current changes occur at an extremely rapid pace, requiring adjustments, strategy modifications, and adaptation to new technologies • This competency is intrinsically linked to observation, active communication, empathy, analysis, and decision-making • Flexibility enables companies to adjust to new competencies and maintain the necessary competitiveness to remain active in the market
10. Achievement to the organization	• It is the drive to achieve challenges, recognize progress made, and grow personally, socially, and professionally • According to Maslow's hierarchy of needs, achieving personal accomplishments is of utmost importance; these achievements should be realistic, attainable, and pursued with excellence in results. When personal goals are met, broader objectives are more likely to be achieved • Achievement motivation should be based on a continuous evaluation of one's personal situation relative to the current professional context, while developing the aspects necessary to reach goals, establishing micro-goals to maintain active motivation
11. Commitment to the organization	• Also known as organizational commitment, it is understood as an affective connection with the organization and the achievement of its goals • The individual must accept the objectives, demonstrate willingness to contribute efforts for the benefit of the organization, and have the desire to remain within it • Three differentiated components: 1. Affective commitment, connection with the organization's values and philosophy; 2. Continuance commitment, attachment of a material nature; 3. Normative commitment, a sense of loyalty driven by cultural or familial pressures
12. Ability to work under pressure	• Stress resistance and self-control • Working under pressure is understood as performing tasks under extreme conditions, with work overload, while maintaining efficiency in results • Self-control over one's own emotions, even when operating under constant stress • The individual should: 1. Identify key objectives and review the work content; 2. Ensure the availability of necessary resources for the assigned task; 3. Confirm objectives and resources with supervisors • This ability provides emotional balance and maintains control over one's emotions, preventing negative reactions that could provoke opposition or hostility from other employees under stressful conditions

(continued)

Table 2.1 (continued)

Skills	Characteristics
13. Conflict resolution and negotiation techniques	• This competency can be learned by developing an appropriate process, such as identifying, defining, analyzing the problem, and seeking suitable solutions and alternatives for all parties • Managing negotiation involves understanding the four possible outcomes in a negotiation: 1. Lose-lose, where neither party achieves their interests or needs; 2. Win-lose, where only one party achieves its objectives; 3. Win–win, where both parties obtain positive results; 4. No outcome, where communication was not successfully established • Negotiation is the tool that enables conflict resolution
14. Pursuit of excellence	• It is the acquisition of habits that lead an individual to perform tasks in the best possible way, without competing with others, as everything is personal • According to Hassan (2006), "excellence has no speed limits, and for this, freedom, decision, will, and intelligence are required." • This skill, aimed at achieving and maintaining excellence, requires the continuous and deliberate improvement of all business activities, including research, design, marketing, finance, production, logistical support, distribution, and sales
15. Innovation	• Innovation is the outcome of activities through which an element is introduced with the intention of benefiting society or the organization; it must involve a change or challenge in conditions, resulting in a new practice or novel application • The ability to generate ideas, develop them, evaluate them based on feasibility criteria, and implement them as solutions to proposed problems or improvements in any professional field
16. Entrepreneurship	• It refers to an individual who initiates a new project, whether as an employee or entrepreneur, making an additional effort to achieve a goal or objective • Within organizations, it is the capacity to drive changes in production models • The personal characteristics associated with this skill include seeking opportunities, passion for initiative, persistence, adherence to work commitments, demanding efficiency and quality, willingness to take calculated risks, setting goals, planning, and executing the necessary follow-ups

(continued)

Table 2.1 (continued)

Skills	Characteristics
17. Leadership	• Personal capacity to act as a support, indicator, or guide for others, influencing behavior and inspiring actions in their surroundings • According to Maxwell (2007), in his book "*The 21 Irrefutable Laws of Leadership*", leadership represents the ability to enhance the performance of individuals within an area through the guidance or direction of a leader • A leader is a person in continuous personal development, who inspires, motivates, and guides individuals or groups, becoming a role model through positivity, interpersonal skills, simplicity, truthfulness, idea transmission, adherence to codes or principles, empathy, charisma, among others • Leadership is the capacity to leverage both flaws and strengths in order to transform individuals, create new opportunities for success, and influence economic and social contexts • Leadership is increasingly human-centered, requiring enhanced communication skills and a profound understanding of the perspectives of each member within a workgroup

Simulation facilitates the transfer of theoretical skills to real-life contexts, giving students the opportunity to refine their performance and gain confidence in their abilities.

5. Feedback: This is one of the most important points, as it constitutes a crucial step in the skill acquisition process. Feedback is provided through supervision by instructors or self-assessment, where performance is observed and evaluated against the desired standard.

The way in which soft skills are acquired, when they are not innate, is due to the necessity to develop the individual in order to perform an assigned or required task. Figure 2.2 presents the model of soft skills learning.

Soft skills emerged from the need to hire personnel who are not only technically trained but also prepared and motivated to perform tasks. These competencies serve as the driving force of leadership, as their traits often have a greater influence than intellectual capacity. Goleman [4], in his book *Emotional Intelligence*, throughout his research and explanations, emphasizes on SS for the success in life depends 80% on personal emotions and only 20% on knowledge acquired or developed through experience. Within 80%, soft skills, including leadership, play a fundamental role.

At universities, it is common to observe individuals with extensive knowledge, doctoral degrees, high levels of specialization in certain fields, and broad professional experience, yet who lack the ability to work in teams, act assertively, or effectively transmit knowledge to students. For this reason, the World Economic Forum established that by 2025, 50% of professionals will be required to possess

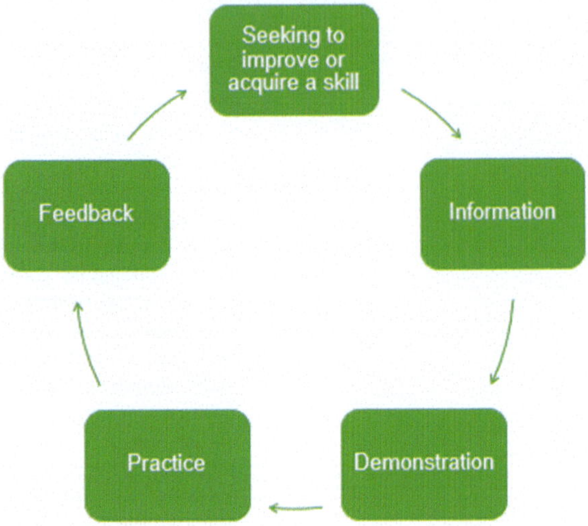

Fig. 2.2 Soft skills learning model. Created by the author

soft skills to achieve workplace success. *The Future of Jobs Report 2020* provided a list of in-demand soft skills, which include leadership and social influence, critical thinking, analytical and innovative thinking, active learning strategies, complex problem solving, creativity and initiative, resilience, and flexibility.

According to the Ministry of Public Education, soft skills are increasingly present within educational curricula and study plans. However, they are not systematically incorporated into the engineering programs offered within the National Technological System. While continuous improvement has been pursued through modifications in curricula, the technical core remains dominant, and concrete actions to develop soft skills are rarely implemented. Instead, suggested or optional activities related to certain topics are provided, but these are not always followed by faculty. In practice, most professors prioritize hard skills so that students achieve knowledge acquisition, leaving the development of soft skills to the students themselves. In many cases, this development does not occur due to the lack of systematic monitoring.

Today, universities recognize soft skills as highly relevant in the training of professionals capable of addressing problems and needs within companies or organizations where engineering graduates will eventually work. These competencies allow engineering graduates to maintain continuous updating, ensuring favorable labor mobility not only for them as individuals—by increasing their employability through diverse abilities—but also for the organizations they serve, particularly in the development of projects and the achievement of objectives. Such accomplishments largely depend on teamwork, leadership, communication, and effective group management.

Currently, companies acknowledge that soft skills represent the true value of their human capital and have begun to view them as factors that foster business growth. Through collaborative channels of communication between companies and universities, employers increasingly demand that engineering graduates not only master technical knowledge but also arrive with well-developed skills that enable them to socialize, organize, lead, and communicate corporate objectives to their peers. Engineering, therefore, is not considered an isolated discipline but rather the operational center of most companies. Consequently, soft skills have become central in personnel selection processes, as they are acquired through professional and life experiences and are considered crucial in determining workplace success. The development of a soft skill is influenced by multiple factors. From early childhood, interactions with family and close environments play a crucial role in shaping competencies such as communication, teamwork, and empathy. Fundamental values, conflict resolution abilities, and respect for others are learned through observation and practice within family and community contexts.

Although soft skills are not typically taught explicitly in formal education, they are developed through collaborative activities, problem-solving, and participation in projects. Additionally, extracurricular experiences such as sports, volunteering, or membership in clubs foster competencies like adaptability and leadership. In both personal and professional domains, challenges strengthen competencies such as resilience, decision-making, and conflict resolution. Difficult experiences can cultivate patience, emotional regulation, and the capacity to work under pressure. The development of soft skills also requires continuous self-reflection and self-awareness. As individuals mature, they learn to manage emotions more effectively, communicate with greater clarity, and empathize with others, thereby promoting personal and professional growth. Observing and learning from influential figures such as mentors, leaders, or peers is also essential for reinforcing these abilities. The feedback and support received from these key figures are fundamental for ongoing development in this field.

Based on practical experience, it can be concluded that soft skills originate and are refined through social interaction and self-exploration, becoming stronger through consistent practice across various life contexts.

2.2 Hard Skills (HS)

A hard skill originates through a structured learning process and develops by acquiring theoretical and practical knowledge in a specific area or environment. These are technical competencies that can be measured and evaluated objectively, and their development follows a more formal and methodical path. After the Cold War, in the mid-1990s, companies realized that hard skills alone were no longer sufficient. Beyond simply employing personnel capable of performing tasks, organizations required individuals who would commit to projects, thereby ensuring solid and global expansion.

The behavioral characteristics of individuals with highly developed hard skills include the following: receiving orders and solving problems immediately; being competent in only one specific area; willingness to remain in the same field for an indefinite period; possessing academic knowledge and professional experience; and demonstrating memory, logic, language proficiency, among other abilities. People who exhibited these characteristics are today identified as the Baby Boomer generation. Their primary goal was retirement, with the expectation of being secure and comfortable until the end of their lives. They focused exclusively on fulfilling their workday and meeting production quotas without emotionally engaging with colleagues or subordinates. Their main motivation revolved around economic security, consumer capacity, and competitiveness, in other words, doing what needed to be done even before superiors requested it.

In essence, hard skills encompass all the technical knowledge, competencies, and abilities that individuals acquire through formal education, training, or life experience, enabling them to perform a specific task or job within a company or organization. Examples include the use of technology, language proficiency, technical design, project management and administration, accounting and finance, as well as the arts, among others. These skills are learned and reinforced not only in schools and universities but also throughout professional and personal life. The following figure illustrates some of the hard skills that are commonly acquired in educational institutions (Fig. 2.3).

These skills remain valuable in companies, but they only represent 50% of success when applying for a job. The other 50% lies in soft skills. For college students, it's vitally important to develop their knowledge and social skills comprehensively.

Fig. 2.3 Hard skills. *Source* Gisela Haro Esquivel [5], images taken from the internet

2.3 Labor Insertion (LI)

Job placement or labor insertion (LI) occurs when an individual begins, for the first time, the transaction of knowledge and skills in exchange for financial remuneration. However, finding a job that fully meets personal needs is a significant challenge faced by graduates or students of IELMEC as they search for professional residencies in the local industry (Ciudad Juárez). Local industries publish vacancies daily, allowing students and engineers to apply for specific positions. Nevertheless, it has become increasingly evident that companies seek not only technically skilled personnel with the necessary hard skills to perform specific tasks, but also individuals who possess so-called "soft skills." For this reason, it is important to explore the relationship between soft skills and job placement. Researching some of the most well-known agencies today, such as OCC, Indeed, and *Empleos Maquila*, reveals interesting data from their publications. For example, Indeed explains objectively what job placement entails and, through its editorial team in June 2023, briefly presented its benefits. According to their publication, the first job may not always meet the individual's expectations, but it is necessary to begin, even if it does not fully align with the applicant's requirements. Therefore, job placement, as the term suggests, refers to becoming incorporated or integrated into the productive sector through employment. Despite the continuous availability of job opportunities, obtaining the desired position remains difficult, whether due to disagreements regarding financial compensation or the lack of corporate training programs. An adequate job placement strategy would, therefore, ensure access to dignified employment that benefits both the worker and the company.

Some of the benefits presented by Indeed include the following: excluded individuals or people with disabilities acquire a sense of belonging and commitment by feeling part of a team; workers contribute to improving the workplace environment with optimism; adaptation to the work environment becomes easier and requires less time; socially responsible companies gain greater recognition from consumers; and job placement programs objectively contribute to improving quality of life. Such programs are essential to reducing discrimination, thereby enabling more people to access decent jobs with fair remuneration. It must be noted that barriers may be physical, such as mobility limitations or illnesses, or more subjective, such as ethnic, religious, or gender discrimination.

Universities play a fundamental role in the job placement of students and recent graduates by constantly updating their curricula to provide graduates with the tools required by the industrial sector. This capacity is crucial for ensuring that students gain greater advantages that support their future professional success. Regarding this desired success, Espinoza [6], in his article *Soft Skills in Education and Business: A Systematic Mapping*, asserts that it depends on the skills and techniques acquired by students or graduates during their time at university. One of the aspects where students should receive support is in job interviews, as applicants have increasingly realized that most companies in the region require soft skills in their job postings. This, in turn, facilitates faster and more satisfactory job placement. To achieve this, Ayala [7] emphasizes that there must be freedom in education, allowing diverse scenarios through which teaching processes occur and where so-called "invisible skills" are naturally developed. These skills, although not always explicit, are essential to facilitating learning processes.

Engineering graduates are among those with the highest job placement rates. According to statistics from the Polytechnic University of Madrid (UPM), more than 90% of its graduates quickly find employment, with nearly 100% eventually entering the corporate sector. In contrast, at the National Technological University (UTN) in Buenos Aires, reports indicate that graduates in Computer Systems Engineering achieve the highest job placement rate at 35%, followed by Industrial Engineering at 15% and Mechanical Engineering at 13%. In Mexico, according to the National Institute of Statistics and Geography [8], the *National Survey of Occupation and Employment* for the first quarter of 2023 presents the following figures by degree program, as shown in Table 2.2.

Thus, it can be observed that graduates of Electromechanical Engineering, whose employment percentage is very similar to that of Mechanical Engineering graduates, enter the labor sector with an approximate starting salary of $15,450 MXN. This indicates that graduates of this discipline receive remuneration above the national average, which directly impacts personal satisfaction. However, an important question arises regarding the extent to which soft skills have contributed to the successful integration of these graduates into the local productive sector, as well as the degree to which such skills have facilitated access to better-paid positions.

Table 2.2 Updated figures for the first quarter of 2023 from the National Survey of Occupation and Employment

Carrera/carrer	Profesionista s Ocupados/ employed professionals	Hombre s/ men (%)	Mujeres/ women (%)	Ingreso mensual promedio ($)/ average monthly income (MXN $)
Ciencias de la computación/ computer science	747,005	71.8	28.2	$14,876
Construcción e ingeniería civil/ construction and civil engineering	242,491	88.4	11.6	$15,264
Electricidad y generación de energía/electricity and power generation	72,773	92.5	7.5	$16,169
Electrónica y tecnología de telecomunicaciones/electronics and telecommunications technology	317,550	93.9	6.1	$15,794

(continued)

Table 2.2 (continued)

Carrera/carrer	Profesionista s Ocupados/ employed professionals	Hombre s/ men (%)	Mujeres/ women (%)	Ingreso mensual promedio ($)/ average monthly income (MXN $)
Industria de la alimentación/ food industry	35,774	43.3	56.7	$11,898
Ingeniería de vehículos de motor, barcos y aeronaves/motor vehicle, ship and aircraft engineering	44,282	97.3	2.7	$13,755
Ingenieríaindustrial, mecánica y metalurgia/industrial, mechanical and metallurgical engineering	138,014	93	7	$15,450
Ingenieríamecánica, electronica y tecnología/mechanical, electronic, and technology engineering	426,194	76.4	23.6	$15,066
Ingeniería química/chemical engineering	203,924	51.5	48.5	$13,791
Manufacturas y procesos, programas multidisciplinarios o generales/manufacturing and processes, multidisciplinary or general programs	7,779	77	23	$15,676
Minería y extracción/mining and extraction	32,815	81.5	18.5	$16,943
Pesca/fisheries	5,111	91.1	8.9	$12,180
Producción y explotación agrícola y ganadera/agricultural and livestock production and exploitation	160,653	88.8	11.2	$13,509
Servicios de transporte/ transportation services	11,402	90.5	9.5	$20,498
Silvicultura/forestry	5,087	79.7	20.3	$14,692
Tecnología de la información y la comunicación/information and communication technology	44,430	79.2	20.8	$16,482
Tecnología y protección del medio ambiente/environmental technology and protection	22,453	50	50	$11,183

Source STPS–INEGI

References

1. Robbins, T.: Soft Skills and Success in Working Life | Ideas Pearson (pearsonlatam.com) (2023). https://blog.pearsonlatam.com/talento-humano/habilidades-blandas-exito-laboral
2. Goleman, D.: In: Kairós, S.A. (eds.) Social Intelligence: the New Science of Human Relationships, 1st edn, Mexico (Essay) (2011)
3. Pita Yañez, C.: How to be competent, professional competences demanded in the labor market. In: Professional Insertion Chair Caja Rural Salamanca (ed.). University of Salamanca, Spain (2013)
4. Goleman, D.: Emotional Intelligence: why it is More Important Than IQ/Emotional Intelligence. B de Bolsillo. USA. Kindle Edition (2018)
5. Haro Esquivel, G., & Ayala Hernández, P.: Mechatronic Engineering: soft Skills for Labor Insertion. Spanish Academic Publishing (2024)
6. Espinoza Mina, M.A., Gallegos Barzola, D.: Soft skills in education and business: systematic mapping. UISRAEL Sci. J. (Universidad Israel, Quito, Ecuador) (2020)
7. Ayala Hernández, P.: Teaching with freedom in higher education in Mexico. Nóesis: J. Soc. Sci. Hum. **22**(44), 44–61 (2013). ISSN 0188-9834. https://dialnet.unirioja.es/servlet/articulo?codigo=5094972
8. National Institute of Statistics and Geography (INEGI).: National Survey of Occupation and Employment (ENOE). Statistics of Professionals in Mexico (2020). https://www.inegi.org.mx/contenidos/saladeprensa/aproposito/2020/EAP_EconFin.pdf

Chapter 3
Methodological Process

This chapter outlines the methodological process followed in the development of the research. It begins with a description of the selected methodology, justifying its relevance to the study objectives. The chosen research approach is then presented, explaining how it guides the analysis of the relationship between soft skills and labor market integration.

Subsequently, the chapter details the data collection instruments used to obtain reliable and valid information from participants. Special attention is given to the design and application of validated questionnaires for soft skills (SS), which serve as the primary tool for measuring students' competencies and their alignment with labor market demands. This methodological framework ensures rigor, coherence, and consistency throughout the investigation.

3.1 Methodology

The methodological perspective in qualitative research, according to Denzin and Lincoln [1], is an approach that focuses on understanding and exploring social and human phenomena from the subjectivity of the participants, allowing for an in-depth comprehension. Through flexible and interpretive methods, researchers are able to capture the richness and diversity of social and cultural realities, providing a profound and detailed vision of the phenomena under study.

In contrast to quantitative research, Creswell and Poth [2] state that while the latter emphasizes measurement and generalization of results through statistical methods, qualitative research relies on the collection of non-numerical data, such as observations, interviews, and document analysis, with the purpose of capturing the complexity and richness of the phenomena being studied.

G. H. Esquivel and P. A. Hernández, *Bridging the Gap: Comprehensive Project Design for Soft Skills Development and Labor Market Integration*, Information Systems Engineering and Management 82, https://doi.org/10.1007/978-3-032-19026-0_3

Qualitative research is characterized by its inductive orientation, meaning that the researcher does not begin with predetermined hypotheses, but rather allows themes and patterns to emerge from the data collected during the research process [3].

This provides flexibility to explore new ideas and perspectives, as well as to adapt the research design as the study develops.

The most common qualitative methods include in-depth interviews, focus groups, participant observation, and content analysis, as mentioned by Merriam [4]. These methods enable researchers to immerse themselves in the social and cultural contexts of the participants, facilitating the capture of the complexity and diversity of human experiences.

According to Lincoln and Guba [5], in qualitative research, validity and reliability are approached differently than in quantitative research. Instead of focusing on replication and generalization of results, qualitative inquiry seeks credibility, transferability, dependability, and confirmability. These criteria emphasize trustworthiness in the data and the interpretation of findings, as well as transparency and coherence throughout the research process.

3.2 The Approach

This research was conducted using a qualitative approach, employing a descriptive and comparative cross-sectional design. The study included a group of individuals in the final semesters of their degree program who were already employed, as well as a group of graduates who shared their experiences of going through the recruitment process for the first time through an interview. Participants were interviewed and asked to complete a questionnaire designed to collect data, with the purpose of examining the potential relationship between soft skills and employability. It is worth noting that the instrument was administered to active students, since it is common for them to begin working during their professional residency prior to graduation. In fact, those who wait until after graduation to enter the labor market represent a minimal number.

To achieve this objective, previously validated questionnaires available in various sources such as Redalyc, Scielo, Sciendo, SSRN, as well as instruments used by research firms, were reviewed. Based on this review, a specific instrument for this research was developed, which was subsequently evaluated and validated by three human resources experts with professional experience in personnel selection for companies.

These experts, belonging to the Department of Economic-Administrative Sciences at the Technological Institute of Ciudad Juárez (ITCJ), hold master's and/or doctoral degrees in Business Administration and are professionally active in the field. Their role was to ensure that the instrument was coherent, clear, relevant, and sufficient, as stated by Escobar and Cuervo [6] in their book *Content Validity and Expert Judgment*. Additionally, the instrument underwent Cronbach's Alpha validation to assess its reliability.

According to Smith [7], soft skills, also known as social or emotional skills, were considered crucial for success in the contemporary work environment, encompassing aspects such as effective communication, teamwork, problem-solving, and adaptability.

On the other hand, Jones et al. [8] defined employability as the process through which individuals manage to access satisfactory employment that matches their skills, interests, and professional expectations.

In this type of study, a focus group of individuals entering the labor market or in the early stages of their professional careers was selected. These participants were assessed in terms of their soft skills using validated instruments, such as questionnaires or structured interviews, which measured aspects like effective communication, teamwork, problem-solving, and adaptability [7].

Subsequently, data related to employability were collected, including employment rate, job stability, job satisfaction, and the level of success in professional roles. These data were compared among individuals with varying levels of soft skills to determine whether a relationship existed between these skills and employability [8].

The descriptive design allowed the identification of the sample characteristics in terms of soft skills and employment status, while the comparative design facilitated the comparison between groups of individuals with different levels of soft skills in order to analyze how these skills influenced their integration into the labor market.

This type of study provides relevant information about the importance of soft skills in professional success and can guide the design of skill development programs aimed at improving employability and adaptation of individuals within the labor market—programs that could be designed in a subsequent study or further research.

3.3 Data Collection Instruments

This chapter focuses on the use of interviews as a key instrument for data collection, offering qualitative insights into the development of soft skills (SS) and their influence on labor market insertion (LI). Unlike standardized tools such as questionnaires, interviews provide the opportunity to explore individual perspectives and lived experiences in greater depth, capturing the nuances of how engineering students and graduates perceive the importance of these competencies in their professional growth.

Through the analysis of participants' responses, this chapter examines both the role of soft skills—such as communication, adaptability, leadership, and teamwork—and the challenges and opportunities associated with labor market integration. The information gathered highlights the connection between academic training and professional practice, offering a more comprehensive understanding of the research problem and complementing the quantitative findings of the study.

3.3.1 Validated Questionnaires for Soft Skills (SS)

The questionnaires validated by leading experts in soft skills provide the opportunity to analyze the specific skills that are intended to be fostered. In this research, the focus is oriented toward the skills developed during professional training in engineering. According to Donamis [9], a leading organization in human resources and recruitment, soft skills are defined based on three key elements: personal skills, social skills, and career-related personal attributes. These elements can be applied to any context in which an individual operates, ranging from personal to professional domains, without restriction, since they are life characteristics. Some of the questions suggested by the organization to evaluate these skills during interviews for potential recruitment include:

1. Questions to assess adaptability
   - Has your organization ever introduced a change with which you disagreed? Why, and how did you adapt to the change?
   - Organizational policies sometimes change. Have you ever faced such a change? How did you respond?
   - In case you have ever changed jobs, how did you manage it? What difficulties did you encounter? Please describe your personal process of adaptation.
   - Organizations are generally progressing. Could you describe how your position has changed? How did you handle it?
   - Could you describe a recent situation in which you had to adjust your plan or rethink your priorities? What was the context and what exactly did you do? What adjustments did you make and how did you implement them? What was difficult for you to let go?
2. Questions to assess ambition
   - What would motivate you to change jobs?
   - How do you evaluate your progress in achieving your set goals?
   - Tell me about a decision you made to advance in your professional career.
   - What does professional success mean to you?
   - What quality do you lack that you would like to have?
3. Questions to assess creativity
   - How would people communicate in a perfect world?
   - Could you tell me about a project in which you deviated from conventional paths and processes? What result did you achieve? What was the reaction of your colleagues/clients?
   - Sell me a trip to space.
   - What is your most creative idea, and why?
   - If you could have a superpower, what would it be ?

4. Questions to assess stress management
   - Tell me about a stressful situation in your professional life. How did you handle it?
   - What stresses you the most at work?
   - How do you avoid stressful situations?
   - What do you do to reduce or mitigate stress?
   - In what way does stress help or affect you at work?
5. Questions to assess leadership
   - If you feel that the work of a team member you are leading is not very effective, what do you do?
   - With whom do you prefer to work? What characterizes that relationship?
   - What superhero would you be?
   - How do you foster the growth of your collaborators' skills?
   - Tell me about a situation in which you took the lead on a project.
6. Questions to assess critical thinking
   - Why do you want to change jobs?
   - What would need to change in your current company for you to stay?
   - What reservations do you have about joining our company and team? What makes you hesitate?
   - What is the first thing you would change about our company?
   - What do you think about our website?

The Secretariat of Public Education [10] designed an instrument to assess social and emotional skills, aiming to identify the level of importance that students attribute to each of the categories presented. The evaluation is left to the teacher's perspective, who determines the applicable aspects for each student under their responsibility. This instrument does not specify the area or region of application, the student's social background, parents' educational level, or any deeper contextual analysis. It is important to note that this evaluation instrument is designed for basic education (kindergarten, elementary, and secondary school). Table 3.1 the image obtained from this instrument is presented.

From this instrument for evaluating social and emotional skills, the method for collecting personal data from potential interviewees was derived, since such information must be kept under strict confidentiality.

There are universities and organizations that have developed their own questionnaires to determine the achievement of transversal or generic competencies, such as oral and written communication, critical thinking, problem-solving, and the development of interpersonal relationships. These competencies represent the soft skills addressed in this research. The questionnaire presented (Table 3.2) indicates whether the individual has control over their emotions, demonstrates leadership, and is accustomed to working under a framework of timely delivery. From this questionnaire, several questions related to emotional control were taken.

For the purposes of that research, a Likert scale was employed, which allowed the results obtained to be visually represented, thereby facilitating the analysis of responses. The use of this scale was key to capturing the individual perceptions of the respondents, given that these may vary significantly from person to person, thus enriching the analysis and providing greater depth to the conclusions.

The Likert scale not only helps visualize the responses but also allows the identification of behavioral and attitudinal patterns among employees. This variability in individual perceptions enables a more detailed exploration of how soft skills, promoted through coaching, can have a differentiated impact on various aspects of the workplace environment, Table 3.3.

Table 3.1 Social and emotional skills assessment instrument *Source* SEP

SEP
SECRETARÍA DE EDUCACIÓN PÚBLICA

Subsecretaría de Educación Básica
Dirección General de Desarrollo de la Gestión Educativa
Programa Nacional de Convivencia Escolar

Instrumento de valoración de habilidades sociales y emocionales

INSTRUCCIONES GENERALES

Previo a la aplicación del instrumento:

1. Revise en la Ficha de Identificación del Instrumento los seis ejes que se abordan.
2. Revise los indicadores que se presentan para los seis ejes, estos ayudarán a detectar las habilidades sociales y emocionales de los alumnos.
3. Aplique la herramienta "*Instrumento de valoración de habilidades sociales y emocionales*" a los alumnos que haya seleccionado previamente.

Al inicio de la aplicación:

4. Recuerde que esta actividad forma parte de un reconocimiento con fines de mejora escolar en los alumnos. NO es un diagnóstico psicológico.
5. No es necesario que los alumnos estén presentes en la aplicación.

Durante la aplicación:

6. Aplique un instrumento por cada alumno seleccionado a partir de lo que ha observado en él. La duración aproximada por instrumento es de 15 minutos.
7. Lea cada uno de los indicadores por cada eje y señale la opción que usted considere que caracteriza al alumno. A partir de la suma de los indicadores señalados por cada eje, ubique el nivel que corresponda.
8. Si duda en seleccionar algún indicador, reflexione respecto a si es algo en lo que sería importante fortalecerlo.
9. Una vez terminados los ejes, registre en la tabla de resumen y realice el gráfico de los resultados.

Fecha de aplicación: ____________________

Nivel educativo:	Preescolar ( )	Primaria ( )	Secundaria ( )	
Modalidad:	Multigrado ( )	Indígena ( )	Educación especial ( )	Otra: ______________
Entidad:			**Localidad:**	
Nombre de la escuela:				**CCT:**
Nombre del Docente:				
Ha participado en el Programa Nacional de Convivencia Escolar: Sí ( ) No ( ) ¿En qué ciclo escolar?				

(continued)

Table 3.1 (continued)

SEP
SECRETARÍA DE EDUCACIÓN PÚBLICA

Subsecretaría de Educación Básica
Dirección General de Desarrollo de la Gestión Educativa
Programa Nacional de Convivencia Escolar

Marque con una ✓ el indicador que generalmente aplique al alumno o alumna.

Posteriormente ubique el nivel que corresponda de acuerdo con el número de indicadores obtenidos.

Nombre del alumno(a): ______________________ **Grupo y grado:** __________

EJE 1. AUTOESTIMA.

( ) 1. Expresa claramente sus ideas y necesidades.
( ) 2. Cuida su apariencia personal.
( ) 3. Se respeta a sí mismo.
( ) 4. Reconoce sus fortalezas.
( ) 5. Manifiesta actitudes de autocuidado.
( ) 6. Mantiene contacto visual al interactuar con los demás.
( ) 7. Se preocupa por los demás.
( ) 8. Reconoce sus áreas de oportunidad.
( ) 9. Ninguna de las anteriores.

EJE 2. MANEJO DE EMOCIONES

( ) 1. Expresa cómo se siente ante diferentes situaciones.
( ) 2. Regula la expresión de sus emociones.
( ) 3. Actúa en forma asertiva en situaciones que le enojan.
( ) 4. Tiene tolerancia ante situaciones que lo frustran.
( ) 5. No tiene fluctuaciones constantes en sus estados de ánimo.
( ) 6. Reconoce y valida las emociones de los demás.
( ) 7. Está regularmente de buen humor.
( ) 8. Es congruente con lo que siente y expresa.
( ) 9. Ninguna de las anteriores.

EJE 3. CONVIVENCIA

( ) 1. Respeta diferentes formas de pensar y/o actuar de los demás.
( ) 2. Convive respetuosamente con sus compañeros.
( ) 3. Se dirige a los demás por su nombre y no por apodos.
( ) 4. Se integra fácilmente a las actividades con otros.
( ) 5. Trabaja en equipo adecuadamente.
( ) 6. Comparte sus cosas.
( ) 7. Ayuda a compañeros cuando lo requieren.
( ) 8. Avisa a una autoridad en el momento que agreden a alguien.
( ) 9. Ninguna de las anteriores.

EJE 4. REGLAS Y ACUERDOS

( ) 1. Cumple la mayoría de reglas y acuerdos del salón y la escuela.
( ) 2. Acepta la redirección del maestro.
( ) 3. Sigue reglas y acuerdos, aun cuando no lo están observando.
( ) 4. Expresa asertivamente desacuerdo con alguna regla o acuerdo.
( ) 5. Es capaz de llegar a acuerdos en forma adecuada.
( ) 6. Escucha diferentes opiniones para elaborar un acuerdo.
( ) 7. Propone reglas o acuerdos de convivencia en el salón o escuela.
( ) 8. Respeta y cede ante reglas y acuerdos a pesar de no estar de acuerdo con ellos.
( ) 9. Ninguna de las anteriores.

EJE 5. RESOLUCIÓN DE CONFLICTOS

( ) 1. Escucha con atención y respeto el punto de vista de los demás.
( ) 2. Evita conflictos en la escuela y su entorno.
( ) 3. Se involucra para ayudar a solucionar conflictos dentro del salón y en la escuela.
( ) 4. Es capaz de negociar con otros para solucionar un conflicto.
( ) 5. Recupera fácilmente la calma después de un conflicto.
( ) 6. No apoya a otros o participa en agresiones a otros niños.
( ) 7. Logra solucionar sus conflictos en forma asertiva.
( ) 8. Es respetuoso con las autoridades de la escuela.
( ) 9. Ninguna de las anteriores.

EJE 6. FAMILIAS

El alumno(a) ...
( ) 1. Se refiere a su familia de forma positiva.
( ) 2. Llega a la escuela en buenas condiciones de aliño
( ) 3. Se siente apoyado por su familia

Los padres o tutores...
( ) 4. Preguntan sobre el desempeño del alumno.
( ) 5. Asisten a eventos escolares a los que son convocados.
( ) 6. Acuden en situaciones que se requiere de su apoyo.
( ) 7. Intervienen en forma positiva en la resolución de un conflicto.
( ) 8. Se involucran en sus tareas escolares
( ) 9. Ninguna de las anteriores.

Cuente los indicadores por cada eje y ubique el nivel que corresponda a cada uno:

ESPERADO (De 6 a 8 indicadores)
EN FORTALECIMIENTO (De 3 a 5 indicadores).
REQUIERE APOYO (De 0 a 2 indicadores).

Señale en el cuadro con una (✓) el nivel obtenido en cada uno.

EJES	Nivel: Esperado	Nivel: En fortalecimiento	Nivel: Requiere apoyo
Eje 1: Autoestima			
Eje 2: Emociones			
Eje 3: Convivencia			
Eje 4: Reglas y acuerdos			
Eje 5: Resolución de conflictos			
Eje 6: Familia			

Observaciones generales: ______________________

The questionnaire is a necessary instrument to facilitate data collection and its subsequent analysis by the researcher, in order to determine future strategies based on the results obtained. Therefore, this model will be considered for the design of the measurement instrument in this study to collect data on soft skills (SS) within the selected sample for analysis. The use of interview instruments to examine the employability of individuals entering the field of engineering has been a subject of interest in academic research [12]. These instruments provide an effective means

Table 3.2 Validated questionnaire to assess soft skills. Questionnaire developed by: National Accreditation Commission of Chile (2015), AKADEMIA magazine [11]

University....................School.........................Current Academic year...........

Only mark one alternative according to your personal opinion (HIGHLIGHT or CHECK).

1. **When new situations arise:**
   a) Becomes upset
   b) Handles them calmly
   c) Analyzes them
   d) Ignores them
   e) Proposes alternatives
2. **When facing a critical problem (e.g., death of a family member, debt, eviction, poor grades):**
   a) Becomes completely depressed
   b) Worries
   c) Assimilates it
   d) Complains
   e) Overcomes it
3. **When listening to proposals from other students, relatives, or friends:**
   a) Pays attention
   b) Does not care
   c) Believes in them
   d) Shares ideas
   e) Encourages them further
4. **Do you take the initiative in a work group?**
   a) Sometimes
   b) Always
   c) Does not commit

(continued)

Table 3.2 (continued)

d) Wishes to, but does not

e) Immediately

5. **How do you consider yourself in writing academic topics for your university work?**

a) Excellent writer

b) Makes many mistakes

c) Average, but corrects

d) Writes just enough

e) Acceptable

6. **How do you approach your university assignments?**

a) Plan them

b) Do them as best as possible

c) Delegate them

d) Ask for help

e) Do them just to comply

7. **When asked to carry out projects:**

a) Assume responsibility

b) It is not my strength

c) Attempt them

d) With little enthusiasm

e) Do not do them

8. **How do you sincerely consider your command of the English language?**

a) Advanced level

b) Intermediate level

c) Basic level

d) Pre-basic level

e) None

(continued)

Table 3.2 (continued)

9. **Regarding knowledge, use, and application of ICT and online learning platforms (Moodle, Classroom, others), how do you self-assess?**
   a) Very good
   b) Good
   c) Fair
   d) Poor
   e) Almost none
10. **Do you have your own opinion about current events (economic crisis, pandemic, government policies, world affairs)?**
   a) Adequately
   b) Very superficial
   c) Indifferent
   d) Interested
   e) No opinion
11. **Do you consider it necessary to pursue additional training beyond the degree you are currently studying?**
   a) Indispensable
   b) If I have the opportunity
   c) Unlikely
   d) Would consider it later
   e) Only the profession I am currently training for
12. **How do you consider environmental education?**
   a) Very important
   b) Important
   c) Moderately important
   d) Slightly important
   e) Not important

Table 3.3 Questionnaire on: (variables) coaching and soft skills survey using a five-point likert scale for the variable *internal control*

Code	Description	Value
A	Always	5
AA	Almost always	4
ST	Sometimes	3
AN	Almost never	2
N	Never	1

No.	Dimension 1: Communication	5 A	4 AA	3 ST	2 AN	1 N
1	Do you consider that ideas improve through teamwork in the area?					
2	Do you consider that appropriate ideas are generated through the application of coaching?					
3	Do you consider that knowledge increases significantly when working in a team?					
4	Do you consider that knowledge is enhanced through the application of coaching?					
5	Do you consider that information is managed more clearly and comprehensively when working in a team?					
6	Do you consider that information is managed more effectively through the application of coaching?					
	Dimension 2: Management	5 A	4 AA	3 ST	2 AN	1 N
7	Do you consider that proper planning of activities improves time management?					
8	Do you consider that activity planning improves through the application of Coaching?					
9	Do you consider that proper execution of tasks contributes to better time management?					
10	Do you believe that tasks are executed more effectively through the application of coaching?					
11	Do you believe that being organized helps improve the response time management of the area?					
12	Do you consider that information is managed more effectively through the application of coaching?					
	Dimension 3: Job satisfaction	5 A	4 AA	3 ST	2 AN	1 N
13	Do you consider that over time your productivity has improved based on decision-making?					
14	Do you consider that productivity increases significantly through the application of coaching?					
15	Do you consider that quality work is achieved when good decision-making exists?					
16	Do you consider that work quality improves through the application of coaching?					
17	Do you consider that team profitability improves with sound decision-making?					

(continued)

Table 3.3 (continued)

	Dimension 4: Teamwork	5 A	4 AA	3 ST	2 AN	1 N
18	Do you consider that team objectives improve with fluid communication?					
19	Do you consider that better objectives are achieved through the development of soft skills?					
20	Do you consider that innovation is generated through assertive communication?					
21	Do you consider that innovation is generated when employees develop their soft skills?					
22	Do you consider that coordination is fluid and precise when applying good communication?					
23	Do you consider that good coordination of activities occurs with the development of soft skills?					
	Dimension 5: Time management	5 A	4 AA	3 ST	2 AN	1 N
24	Do you consider that optimizing processes helps improve management?					
25	Do you consider that processes can be optimized through the application of employees' soft skills?					
26	Do you consider that effective use of time improves with team activity management?					
27	Do you consider that proper time management occurs through the application of soft skills?					
28	Do you consider that proper use of resources occurs through good management?					
	Dimension 6: Decision-making	5 A	4 AA	3 ST	2 AN	1 N
29	Do you consider that problem-solving improves when there is job satisfaction within the team?					
30	Do you consider that proper problem-solving occurs through the application of soft skills?					
31	Do you consider that the team has the reasoning capacity to carry out daily tasks and that this improves when there is satisfaction with their own work performance?					
32	Do you consider that better reasoning occurs through the application of soft skills?					
33	Do you consider that the team makes better decisions when job satisfaction exists?					
34	Do you consider that assertive decisions are made through the application of soft skills?					

"We sincerely appreciate your valuable participation."

of delving into the experiences, perceptions, and challenges faced by novice engineers in the labor market. According to Rodríguez (2017), interviews offer a unique opportunity to explore qualitative aspects related to employability, such as professional expectations, adaptation to organizational culture, and the development of technical and soft skills.

One of the key advantages of interview instruments in this context lies in their ability to capture detailed and contextualized information about the employability process in engineering. Through open-ended and semi-structured questions, interviewers can inquire about job search strategies, the factors influencing the selection of the first professional position, and experiences during the first months of employment [13]. This information is invaluable for understanding the specific challenges that newly graduated engineers face in the labor market.

Nevertheless, it is crucial to acknowledge that the design and application of interview instruments to investigate employability in engineering also present methodological challenges [14]. For instance, the appropriate selection of participants and the assurance of the validity and reliability of the data collected are critical aspects that require meticulous attention. Additionally, interviewers must take into account the potential for response bias, such as social desirability bias or the tendency of participants to provide socially acceptable responses rather than fully honest ones.

Another relevant aspect to consider in the use of interview instruments to examine employability in engineering is the diversity of professional trajectories and workplace contexts within the field [15]. Different specialties, industrial sectors, and types of organizations can significantly influence the employability experience of novice engineers. Therefore, it is essential to design interview instruments that allow exploration of this diversity and capture the complexities of the employability process in engineering.

The instrument proposed to be reviewed by experts and subjected to reliability analysis is presented in Table 3.4. It contains a section for the collection of personal and confidential data (7 questions), a section for obtaining data regarding the SS possessed by the individual (6 questions per soft skill assessed), and a final part addressing data related to the participant's employability (10 questions). It is important to mention that the questions will be interspersed to prevent weighting or bias that may arise on the part of the respondent.

The questions were obtained from various previously validated instruments and applied in other investigations. The questions addressing soft skills were adapted from the book *Development and Validation of the Soft Skills Questionnaire (CHB)* by Díaz-García et al. [16], as well as from the book *The Big Five personality dimensions and job performance: A meta-analysis* by Barrick and Mount [17], among others cited within the instrument.

For the employability questions, books and/or published journals were consulted, aiming to facilitate the reader's process of obtaining employment. Examples include the article *Unraveling the impact of the internet on job search strategies and labor market outcomes: A job seekers' perspective* by Claes et al. [18], and the article published in the *Journal of Business and Psychology* entitled *Unraveling the impact*

Table 3.4 Structured instrument by Gisela Haro Esquivel (2023); reviewed by: José Luis López Galván, Ph.D., Verónica Sansabas, M.Sc. Pablo Ayala Hernández, M.Sc. Ana Margarita Núñez Cortez. Prepared by the author

Structured instrument for the detection of soft skills in students and graduates of IELMEC of the ITCJ and its relationship with job placement.										
INSTRUCTIONS. Mark with an X, the option you must identify with.										
Gender	**F**		**M**		**Other**		**Age**		**Years old**	
Student		**Semester**		**7th.**		**8th.**		**9th.**		**10th.**
Recent Graduate		**Manager**			**Engineer**		**Technical**			
Soft Skills										
1. How do you assess your ability to communicate effectively with your colleagues or team? Question oriented toward effective communication [20]										
2. What actions do you take in situations where you have successfully resolved a conflict in the workplace? Question oriented toward conflict resolution [21]										
3. How do you evaluate your adaptability to changes in the workplace or organizational environment? Question oriented toward adaptability [22]										
4. Can you describe an experience in which you collaborated within a team to achieve a common goal, particularly when working with individuals of different personalities or work styles? How did you experience this process? Question oriented toward teamwork and emotional intelligence [12, 23]										
5. Managing stress and pressure at work, as well as time organization and management, are fundamental. How do you handle these aspects? Question oriented toward organizational commitment, time management, and resilience [24—26]										
6. How do you handle constructive criticism in the workplace? Question oriented toward receptiveness to feedback [27]										
7. How do you keep yourself motivated? Question oriented toward motivation [28]										
8. Have you experienced situations in which you demonstrated the ability to work independently? Question oriented toward autonomy [29]										
9. Do you remain up to date in your professional field, and what actions do you take to pursue continuous development in your work? Question oriented toward lifelong learning [30]										
10. What skills do you consider to be your main strengths in the workplace? Question oriented toward self-awareness [31, 32]										
Labor market integration										
11. How frequently do you effectively communicate your skills and professional experiences during job interviews or in professional settings?										In this section, questions were developed focusing on job search skills [33] decision making [34] adaptability [35] social skills [36] and work-life balance [34]
12. How effective do you consider your abilities to be in searching for and applying for job opportunities that align with your professional profile?										
13. What was your main challenge during the first months in your new job, and how did you address it?										
14. What is your approach to establishing effective relationships with your supervisors and colleagues in your new workplace environment?										
15. When you receive feedback during an interview, how do you respond to it?										
16. Do you feel confident when presenting your résumé and during job interviews?										
17. Have you participated in networking activities or contacted recruiters to enhance your opportunities for labor market integration?										
18. What measures do you take to improve your visibility in the labor market and to stand out as a desirable candidate for employers?										
19. How would you evaluate your capacity to adapt to changes in the labor market and to take advantage of available employment opportunities?										
20. Do you believe the salary you obtained was the most optimal one?										

of the internet on job search strategies and labor market outcomes: A job seekers' perspective by Blau and Petrucci [19], among others cited within the instrument.

After completing the interviews, the data must be transcribed and analyzed for each of the questions. This analysis is presented in the following chapter.

References

1. Denzin, N.K., Lincoln, Y.S.: The SAGE Handbook of Qualitative Research. Sage Publications (2018)
2. Creswell, J.W., Poth, C.N.: Qualitative Inquiry and Research Design: choosing Among Five Approaches. Sage Publications (2018)
3. Patton, M.Q.: Qualitative Research & Evaluation Methods: integrating Theory and Practice. Sage Publications (2015)
4. Merriam, S.B.: Qualitative Research: a Guide to Design and Implementation. Revista Scientific Research an Academic Publisher., San Francisco, CA, Jossey-Bass (2009)
5. Lincoln, Y.S., Guba, E.G.: Naturalistic Inquiry. Sage Publications (1985)
6. Escobar, A.F., Cuervo, M.R.: Design and Validation of Measurement Instruments. Pearson Education, Mexico (2008)
7. Smith, J.: Soft skills in the workplace. J. Hum. Resour. Manag. **15**(2), 123–135 (2010)
8. Jones, A., Smith, B., Brown, C.: The role of soft skills in employment success. J. Workplace Learn. **27**(5), 355–367 (2015)
9. Donamis.: Soft skills: what they are and why they are important in job search. Barcelona, Spain (2022). https://www.doonamis.es/soft-skills-que-son/#:~:text=El%20Dr.,an%C3%A1lisis%20de%20habilidades%20por%20comportamientos
10. Secretariat of Public Education.: Assessment tool for social and emotional skills. SEP, Mexico (2018). https://www.gob.mx/cms/uploads/attachment/file/346154/Instrumento_de_Valoracion_de_Habilidades_Sociales_y_Emocionales.pdf
11. Revista Akademia.: Infusion of skills in the curriculum of higher education: key to the development of advanced human capital, vol. 7, no. 1, pp. 53–73 (2016)
12. Rodríguez, A., et al.: Enhancing teamwork in the workplace: strategies and practices. J. Manag. **32**(1), 78–93 (2017)
13. Smith, J., Brown, C.: Exploring early career engineers' perceptions of job satisfaction. Eng. Manag. J. **31**(2), 78–91 (2019)
14. García, A., et al.: Methodological challenges in studying early career engineers' pathways. Int. J. Eng. Res. Technol. **13**(4), 567–580 (2020)
15. Johnson, R., López, S.: Exploring career trajectories of engineering graduates: a longitudinal study. Eng. Educ. **24**(1), 45–59 (2018)
16. Díaz-García, I., González-Morales, M.G., Córcoles, D.: Development and validation of the soft skills questionnaire (CHB). Psicothema **29**(2), 234–240 (2017)
17. Barrick, M.R., Mount, M.K.: The big five personality dimensions and job performance: a meta-analysis. Pers. Psychol. **44**(1), 1–26 (1991)
18. Claes, R., Beheydt, C., Lemmens, B.: Unraveling the impact of the internet on job search strategies and labor market outcomes: a job seekers' perspective. Internet Res. **15**(1), 56–74 (2005)
19. Blau, G., Petrucci, T., McClendon, J., King, S., McClendon, S.: The challenge of assessing employment success. J. Bus. Psychol. **26**(1), 43–54 (2011)
20. Smith, J.: The importance of effective communication in the workplace. J. Bus. Commun. **42**(3), 320–345 (2015)
21. Jones, A., Brown, C.: Conflict resolution strategies in the workplace. J. Appl. Psychol. **25**(2), 167–184 (2018)

22. García, M., López, S.: Adaptability in the workplace: strategies for success. J. Organ. Behav. **38**(4), 487–502 (2019)
23. López, S., Rodríguez, A.: Emotional intelligence in the workplace: key competencies for success. J. Organ. Behav. **29**(2), 210–225 (2017)
24. Johnson, R., Smith, J.: Stress management techniques for the workplace. J. Occup. Health Psychol. **22**(3), 345–362 (2016)
25. Brown, C., López, S.: Time management techniques for improved productivity. J. Appl. Psychol. **28**(3), 278–295 (2017)
26. García, M., Smith, J.: Building resilience in the workplace: strategies for coping with stress. J. Occup. Health Psychol. **26**(1), 120–135 (2018)
27. López, S., et al.: Constructive criticism and its impact on workplace performance. J. Bus. Ethics **35**(2), 210–225 (2019)
28. Johnson, R., et al.: Motivational strategies for employee engagement. J. Organ. Behav. **33**(4), 456–471 (2019)
29. Rodríguez, A., García, M.: Fostering autonomy in the workplace: strategies for success. J. Appl. Psychol. **31**(2), 234–250 (2018)
30. Smith, J., et al.: Continuous learning strategies for career advancement. J. Career Dev. **34**(3), 345–360 (2017)
31. García, M., Rodríguez, A.: Self-assessment strategies for career development. J. Career Dev. **28**(1), 89–104 (2017)
32. Smith, J., Rodríguez, A.: Effective leadership strategies for workplace success. J. Leaders. Stud. **20**(2), 176–192 (2018)
33. López, S., Rodríguez, A.: Employment search strategies and their impact on job placement. J. Voc. Rehabil. **42**(2), 189–204 (2018)
34. Rodríguez, A., et al.: Work-life balance strategies in the early career: a comparative study. J. Occup. Health Psychol. **25**(1), 89–104 (2019)
35. Smith, J., García, M.: Challenges faced by new employees in the workplace. J. Organ. Behav. **25**(3), 321–336 (2017)
36. Rodríguez, A., Smith, J.: Building effective relationships in the workplace: a longitudinal study. J. Bus. Commun. **35**(1), 120–135 (2019)

Chapter 4
Interviews and Response Analysis

The instrument entitled "*Structured instrument for the detection of soft skills in students and graduates of Electromechanical Engineering at ITCJ and their relationship with labor market integration*", presented in Table 3.4, was administered to a total of 15 students and 15 graduates of the Electromechanical Engineering program at TecNM, ITCJ campus. The following section presents a transcribed response from one student and one graduate; however, for the analysis of each question, the responses of all interviewees were taken into consideration.

The analysis was carried out for each of the questions, and based on this, a consolidated response was generated in order to verify the proposed assumption. This assumption states that "*the development of soft skills in students provides valuable tools that lead employers to favor the hiring of those who demonstrate their use, thereby facilitating their labor market integration into the local industry.*" To illustrate the responses, five interviews are transcribed, in addition to the generalized response. Once this process is completed, it will be possible to validate the aforementioned assumption, thus providing a clearer understanding of the influence of soft skills on the labor market integration of students and graduates.

4.1 Interviewed Participants

The interviewees included 15 students, of whom 12 were male and 3 were female. Regarding their ages, 9 were 22 years old, 2 were 23 years old, and 1 was 25 years old. All were in the ninth and tenth semesters, and all 15 were employed in the maquiladora industry at the engineering level, distributed across the medical, harness, and design sectors, with 5, 4, and 3 students, respectively.

Additionally, 15 graduates were interviewed, of whom 13 were male and 2 were female. Their ages varied: 1 was 25 years old, 4 were 32, 6 were 35, and 4 were 41 or older. Among them, two were recent graduates, six held positions as plant

G. H. Esquivel and P. A. Hernández, *Bridging the Gap: Comprehensive Project Design for Soft Skills Development and Labor Market Integration*, Information Systems Engineering and Management 82, https://doi.org/10.1007/978-3-032-19026-0_4

managers, and seven worked in engineering at the departmental supervision level. The companies where they were employed belonged to the maquiladora industry and specialized in sectors such as design, medical products, textiles, harnesses, and electronics.

In total, 30 individuals were interviewed, providing a significant representation of the relationship between soft skills and the labor market integration of Electromechanical Engineering students from the Technological Institute of Ciudad Juárez (ITCJ).

4.2 Questions and Responses Regarding Soft Skills (SS)

The following section presents the questions and a selection of responses obtained regarding the soft skills of the individuals interviewed. These responses reflect the personal perceptions and experiences of each participant in relation to the competencies assessed. Subsequently, the complete set of responses collected was used to construct a generalized view synthesizing the main findings and observed trends within the group. This approach not only makes it possible to identify common patterns but also highlights key differences in individual perceptions, thereby enriching the overall analysis and facilitating the identification of specific areas for the development and strengthening of these skills.

By generating a generalized response, the aim is to provide a balanced representation that allows for an understanding of soft skills collectively, without overlooking the particularities of each interviewee. Student responses are identified with letters A, B, C, … N, while graduate responses are identified with numbers 1, 2, 3, … n.

In Question 1 (Table 4.1), the communicative competence of electromechanical engineers is explored in the context of their work, since this skill directly influences project success by facilitating interaction with colleagues and the work team.

In Question 2 (Table 4.2), the focus is on soft skills related to conflict or workplace problem management. This skill represents the competence required to identify, confront, and effectively resolve discrepancies and disagreements that may arise during the execution of projects or within the workplace context. According to Robbins and Judge [1], this competence is crucial for maintaining a harmonious and productive work environment, since poorly managed conflicts can hinder progress and negatively affect the quality of outcomes. In engineering, where collaboration and coordination are fundamental to the success of projects, conflict management skills acquire particular importance as they facilitate the mitigation of tensions, the resolution of problems, and the maintenance of focus on common objectives.

In Question 3 (Table 4.3), which refers to adaptability to changes in the engineering context, the focus is on the capacity of professionals to adjust and respond effectively to transformations that may arise in the workplace, in technology, or within project development. This skill is essential given the dynamic nature of engineering, where professionals are constantly confronted with new challenges, technological advancements, and modifications in project requirements. As noted by Ivancevich

Table 4.1 Question 1, soft skills questionnaire

1. How do you assess your ability to communicate effectively with your colleagues or team?
1. Good. I present the problem and possible solutions and ask for other opinions. I have done this since I was a student, because I participated in some projects for ASME competitions, and that helped me to communicate my doubts and objectives, as well as to listen to the people I work with
2. I consider it to be assertive, professional, and respectful, although in my first job I was too quiet
A. Good, but it could be better
13. It is always assertive; I manage to explain my ideas in a clear and compelling way
O. Good, just like that
The conclusion that can be drawn regarding the skill of effective communication with colleagues in the workplace is that communication is generally assertive, clear, and professional. Nevertheless, it is recognized that communication can be improved, and the need to dedicate efforts toward achieving this is emphasized

Source Own elaboration

Table 4.2 Question 2, soft skills questionnaire

2. What actions do you take in situations where you have successfully resolved a conflict in the workplace?
5. On one occasion, senior operators mistreated the new ones. Therefore, a line-by-line meeting was held, explaining why teamwork was important and reminding them that they themselves did not start out knowing 100% of the assembly process
C. In the workshops I coordinate, I always have to mediate between different points of view to avoid confrontations between departments
F. By confronting both parties
11. In some projects within the company, I proposed alternatives to existing problems. When these alternatives were tested, favorable solutions were obtained
8. A discussion arose among the work team, which included the maintenance department, production, and my position as an engineer. Differences of opinion emerged regarding an established process. As the area leader, I mainly sought to understand the perspectives of each department in order to reach an agreement to resolve the situation. Consequently, I assigned tasks that helped all of us as a team and as a work area
In conclusion, the findings regarding the soft skill of conflict management indicate that the necessary abilities have been developed to listen, communicate, and reconcile different points of view, thereby facilitating conflict resolution through consensus-based agreements

Source Own elaboration

and Konopaske [2], adaptability to change involves the ability to anticipate, accept, and proactively manage changes, while seeking innovative and flexible solutions to maintain efficiency and effectiveness at work. Therefore, this question aims to determine whether this soft skill is developed among graduates of Electromechanical Engineering (IELMEC).

Question 4 (Table 4.4) *is oriented toward teamwork*. Effective collaboration in engineering teams not only involves the division of tasks and responsibilities but also open communication, mutual respect, and the ability to adapt to different working

Table 4.3 Question 3, soft skills questionnaire

3. How do you consider your adaptation to changes at work or in the workplace?
G. I find it difficult to adapt to sudden changes, unless they are for the benefit of the team and facilitate operations at the workstations
14. I try to evaluate the positive and negative impact of such changes in order to mitigate risks and take advantage of opportunities
I. By being proactive
L. It takes me a few days to adapt to new changes or challenges, but I always manage to respond satisfactorily
9. I am an adaptable person when it comes to changes, as I enjoy learning more about diverse topics or areas of interest that contribute to my professional growth
The response regarding the soft skill of adaptability to change is that starting in an unfamiliar environment is not simple; rather, it requires a period of adjustment and familiarization with the new workplace context

Source Own elaboration

styles. As Robbins and Judge [1] point out, successful engineering teams foster an environment of trust and support, where members can freely express ideas, receive constructive feedback, and work together toward shared goals.

Question 5 (Table 4.5) is aimed at determining whether respondents have developed the ability to manage stress and pressure at work, a crucial skill for professionals, particularly in the field of IELMEC, where projects are regularly complex

Table 4.4 Question 4, soft skills questionnaire

4. Do you have experience or situation in which you worked as part of a team to achieve a common goal, and how did you experience working with different personalities or work styles?
D. In the change of a conflicting filter assembly process that was solved by adding an extra pair of assemblers, the quality and project teams were called to exchange ideas, and input from the assemblers themselves and the line supervisor was considered
4. We worked as a team on product development; it is a challenge to lead with diverse approaches and personalities. Respect and tolerance of frustration are essential to achieving this
H. At work, I have not yet had such an opportunity, but in school, teamwork was chaotic when trying to combine all the ideas, characters, personalities, and work styles. Sometimes, we had to request extra time to deliver the project
3. At the beginning, it is important to clearly define the objective and each member's role to achieve that particular goal. Examples include projects developed within educational facilities
10. Any day is an opportunity to work with different personalities, but when we implemented changes in plant engineering, people's temperaments surfaced, and it was necessary to harmonize situations so that everyone could work on their part and link them all together to achieve the objective
The conclusion regarding the soft skill of teamwork suggests that dealing with diverse personalities and temperaments within the team can be challenging. However, when it is clear that everyone is focused on achieving a common goal, it becomes easier to overcome individual differences and collaborate for the benefit of the group

Source Own elaboration

and demanding. According to Greenberg and Baron (2008), effective stress management involves the ability to recognize and manage work-related tensions and pressures constructively, while maintaining optimal performance and preserving both emotional and physical well-being.

In many cases, feedback in the workplace is not always received positively. For this reason, *Question 6* (Table 4.6) *seeks to identify how respondents manage criticism.* According to Stone et al. [3], effectively handling constructive criticism in the workplace is crucial for professional development and continuous improvement. When criticism is properly perceived, it can provide valuable guidance for personal growth and work performance. It is essential to adopt a receptive and non-defensive attitude toward criticism, recognizing its potential to foster learning and excellence at work.

Motivation in the workplace is a crucial phenomenon that influences individuals' performance and job satisfaction. According to the Self-Determination Theory proposed by Deci and Ryan [4], intrinsic motivation, driven by interest and inherent satisfaction in the activity, is fundamental for work engagement and the quality of the tasks performed. Furthermore, the Goal-Setting Theory of Locke and Latham [5] highlights the importance of setting specific and challenging goals to improve performance and motivation at work. These approaches emphasize the need to cultivate a work environment that fosters autonomy, personal development, and recognition, which are key elements to maintaining high levels of motivation in the workplace. Thus, *Question 7* (Table 4.7) *provides insight into how IELMEC students and graduates seek to maintain their motivation within companies.*

Table 4.5 Question 5, soft skills questionnaire

5. Stress and pressure management at work, as well as time organization and management, are fundamental. How do you manage yourself?
B. By schedules and timelines, setting short-term goals and deadlines to achieve them
2. It is difficult but not impossible; one must review whether priorities can be redefined to balance workload and obtain results. It is a matter of practice, and difficult to achieve without having gone through various experiences
E. By exercising, which reduces my stress and clears my mind from work
K. I must take time to properly manage my activities within the established hours. Having calendars with dates and times for each activity—at the beginning it was a total disaster, although I already had some experience of delivering final projects in my courses
7. I organize my activities by priority to follow up on each of them, and I currently visit a psychologist once a month to learn stress management techniques and improve my professional and personal life
The response regarding the soft skill of stress, time, and pressure management, as well as organization, implies that it is necessary to set priorities and adapt accordingly. Furthermore, it is essential to take care of both physical and mental well-being to maintain balance during stressful situations, and this skill is primarily acquired through experience

Source Own elaboration

Table 4.6 Question 6, soft skills questionnaire

6. How do you handle constructive criticism in the workplace?
L. Any criticism that helps me grow is accepted and analyzed. Explain to me why you believe I am wrong, and then I act based on what I think. I have done this ever since a female engineer mentioned in a conference at the Tech that it is fine to listen, but the decision to change or to recognize that I am actually wrong is mine
M. Always starting from the perspective that it is not personal and is meant to help me improve
13. There is no constructive criticism; they are areas of opportunity
15. I pay particular attention to improving those negative points and enhancing my performance
A. Regularly, although I have received somewhat harsh criticism and, on some occasions, did not react well. However, I usually prefer to remain calm when someone gives me feedback
The response regarding the soft skill of receiving feedback indicates that criticism may initially be perceived more intensely, but over time and with personal growth, one can learn to accept it and view it as an opportunity for improvement

Source Own elaboration

Table 4.7 Question 7, soft skills questionnaire

7. How do you stay motivated?
D. By visualizing short-term goals and meeting deadlines without sacrificing either quality or personnel
H. I always try to learn something new and manage my time efficiently
O. Continuing to study—at the Tech, you learn to stay motivated to complete projects, pass exams, and still have the energy to spend time with friends
11. I set goals, and within each of them, I give myself moments to rest and then move forward again
8. By keeping my personal and professional goals in mind, surrounding myself with like-minded people, taking quality time for myself, and enjoying time with those close to me
The approach to the soft skill of self-motivation is described as a process that requires constant practice, where it is essential to regularly visualize goals and progress, while also allocating time for recreational activities

Source Own elaboration

Question 8 (Table 4.8) *addresses the ability to work autonomously*, referring to an individual's capacity to plan, organize, and execute tasks independently, without direct supervision. According to Deci and Ryan's [4]. Self-Determination Theory, autonomy is one of the three basic psychological needs that drive intrinsic motivation, along with competence and relatedness. Job autonomy is associated with higher levels of job satisfaction, organizational commitment, and psychological well-being. Moreover, autonomy in the workplace is positively related to creativity and innovation, as it allows individuals to explore new ideas and approaches without excessive external restrictions.

Question 9 (Table 4.9) is focused on continuous professional development, a fundamental process for maintaining relevance and effectiveness in an ever-evolving work environment. Research has shown that the acquisition and application of new knowledge and skills contribute significantly to job performance and the professional

Table 4.8 Question 8, soft skills questionnaire

8. Have you experienced any situation in which you demonstrated the ability to work autonomously?
3. Overtime during the third shift, when a quick model change had to be made due to a lack of material, personnel, or storage
A. Decision making is constantly required in the position, and work is performed with minimal supervision
5. When introducing new production lines
9. When I perform activities that I do not fully understand, I research in order to make modifications and comprehend the task or activity in question. Also, when final projects are carried out at Tech, that is entirely decision-making and completely autonomous work
H. In the company where I currently work, we had a labeling validation problem due to missing procedures, formats, and implementation. I took the initiative to present an idea to my colleagues and supervisors to strengthen the release of incoming new projects as well as improve the established ones. I carried out each step and process and trained my colleagues and the personnel involved in this activity. Today, it is widely used in both plants of the company
Regarding the soft skill of autonomy in decision-making, it is highlighted that this degree of independence may vary. However, when an improvement opportunity is clearly identified, a proposal is formulated and, if feasible, implemented. This dynamic is inherently experienced as a leader, but as a subordinate, it must be subject to evaluation

Source Own elaboration

growth of individuals [6]. Furthermore, ongoing professional development enables practitioners to adapt to technological advancements, shifts in labor market demands, and new regulations, thereby enhancing their employability and their ability to address emerging challenges within their respective fields [7].

Question 10 (Table 4.10) aims to gather information on whether respondents have identified soft skills (SS) that facilitate their integration into companies. These include effective communication, teamwork, problem-solving, empathy, and adaptability [8]. Research has demonstrated that soft skills are important predictors of

Table 4.9 Question 9, soft skills questionnaire

9. Do you stay up to date in your professional field, and what actions do you take to continuously develop at work?
O. Technical journals and social media
3. Always keeping track of available courses related to my field of work,
G. I struggle; sometimes I have to take weekend courses because between school and work, weekdays are not enough
C. I look for courses that provide knowledge in my area of work
7. By researching topics of interest. I am currently pursuing a master's degree in industrial engineering to strengthen my knowledge of the industry and improve my administrative skills within my professional field
When considering the soft skill related to professional updating, it is emphasized that this becomes more attainable by leveraging online networks, as they provide courses available at flexible times, facilitating ongoing training and knowledge updating

Source Own elaboration

Table 4.10 Question 10, soft skills questionnaire

10. What skills do you consider your strengths in the workplace?
2. I learned to work under pressure, keep a cool head, and not take things personally
6. Organized, self-taught, perseverant, firm, and respectful
F. Assertive communication and social media management
M. I can easily perform manual tasks without needing to practice them regularly
11. Leadership, communication skills, digital skills, creativity, and proactivity
Regarding the soft skill of self-awareness, it is emphasized that individuals are aware of their strengths, recognize and utilize them, while also identifying areas that require development, which they apply as necessary

Source Own elaboration

job performance and satisfaction, as they influence individuals' ability to collaborate effectively, manage conflicts, and adapt to changing work environments [9]. Soft skills are considered an invaluable asset that complements hard skills (HS), contributing to long-term professional success.

4.3 Questions and Answers on Labor Insertion (LI)

The following presents the questions and answers obtained in the interviews concerning labor insertion. For each question, a generalized response was formulated to be used more effectively and to encompass most of the responses obtained.

Question 11 (Table 4.11) *aims to gather information on whether respondents consider it essential to be a strong candidate for employment*, depending on the frequency with which they highlight their soft skills and prior work experiences that make them stand out positively. Research indicates that the ability to effectively communicate soft skills, such as teamwork, interpersonal communication, and leadership, increases the likelihood of being considered for a position [10].

Table 4.11 Question 11, soft skills questionnaire, labor insertion section

11. How often do you effectively communicate your skills and work experiences during job interviews or in professional situations?
F. It is difficult because I lose track of the conversation, and in the end, I am not sure whether I said what they wanted to hear or if I just told them a story about my life
I. I just answer what is asked, so I consider it regular
12. I usually communicate very well; I am very clear in my responses
N. When they ask the right questions, the answers flow naturally
15. I always communicate correctly. Since I like to highlight my strengths, it is easy for me to talk about the things I know, and I emphasize even more the things I do well
The perception regarding the importance of communication in labor insertion is that individuals have cultivated communication skills that are fundamental when interacting with potential employers in the workplace

Source Own elaboration

Question 12 (Table 4.12) seeks to determine whether interviewees perceive that searching and applying for job opportunities aligned with their professional profile is a crucial process in career development. According to Self-Determination Theory, individuals tend to seek work environments that satisfy their basic psychological needs, such as competence, autonomy, and relatedness [11]. Thus, the perceived fit between an individual's professional profile and job characteristics influences their motivation to apply for such opportunities [12]. Research has demonstrated that a positive perception of fit increases job satisfaction, organizational commitment, and the likelihood of job retention [13].

The first months in a new job present a series of challenges that can affect adaptation and work performance. The transition into a new work environment may generate stress and anxiety due to the need to learn new responsibilities, establish relationships with colleagues, and adapt to organizational culture [14].

Furthermore, the pressure to demonstrate competence and performance may generate insecurity and fear of failure (Morrison 2002). To address these challenges, individuals often resort to coping strategies such as setting realistic goals, seeking social support within the organization, and developing emotional self-regulation skills [15]. The willingness to learn and proactively adapt to the new demands of the position also plays a key role in overcoming these challenges and achieving professional success in the new role [16]. *Question 13* (Table 4.13) *focuses on obtaining information about how interviewees addressed these new challenges.*

Table 4.12 Question 12, soft skills questionnaire, employability section

12. How effective do you believe your skills are in searching for and applying for job opportunities that align with your professional profile?
C. The size of the company, the salary, and the benefits, in addition to the fact that I really liked the company's prestige
G. Salary
2. I have adjusted quite well to the use of social networks
J. I found it very easy and enjoyable; the work environment is particularly good
M. The challenge of adjusting to a new work team, especially when you join as an engineer and are younger than the maintenance technicians, who want to test your knowledge and subject you to hazing
11. Becoming part of the team, working with people of different ages, social backgrounds, and education levels, who tend to rely on first impressions, is somewhat difficult. However, within two weeks, they realize the way I work, and we are able to reach an agreement that benefits everyone and allows us to meet production expectations
B. Performing different activities outside my area of expertise and managing to solve them through study
5. Adapting to changes in decision-making, dealing with difficult personalities, as well as a heavy workload. The area assigned to me was disorganized at the time and full of opportunities for improvement
The way in which individuals adapt to new work environments is generally characterized as a process that occurs over a short period, during which they become familiar with their colleagues and the workplace environment

Source Own elaboration

Table 4.13 Question 13, soft skills questionnaire, employability section

13. What was your main challenge during the first months in your new job, and how did you address it?
6. Personnel management. Human resources are challenging; managing the personalities of 175 people under my supervision, with different abilities, problems, and ways of thinking, and unifying them while convincing them of the goals to be achieved, is a very difficult task. By taking all the courses and resources provided by the company and with the unconditional support of my direct supervisor, the work always gets done
8. I like to maintain good communication for professional reasons and a polite demeanor, as this makes the environment more pleasant and the workload feel lighter. Personally, I like to keep professional matters separate from my private life at work
D. Respect and fulfilling my work on time and as required; I try not to get
K. Assertive communication
L. Performing correctly within my area of responsibility
The perspective on employability in terms of interpersonal relationships suggests a preference for clearly separating work relationships from personal ones. In the workplace, they consider it essential to establish relationships that contribute to achieving professional goals

Source Own elaboration

Effective relationships with supervisors and colleagues in the workplace are crucial for individual integration and professional success. *Question 14* (Table 4.14) *focuses on this soft skill*, as the quality of such relationships is linked to job satisfaction, organizational commitment, and work performance [17]. Similarly, Wayne et al. [18] states that positive relationships with supervisors and colleagues not only facilitate social and emotional integration into the new work environment but can also provide opportunities for professional development and career advancement.

Question 15 (Table 4.15) *is intended to gather information regarding the management of criticism during a job interview.* The manner in which feedback is received

Table 4.14 Question 14, soft skills questionnaire, employability section

14. What is your approach to establishing effective relationships with your supervisors and colleagues in your new work environment?
F. The unconditional support of my direct supervisor always ensures the work is accomplished, which turns the relationship into a reciprocal one, not only employee–subordinate but also colleagues working together
3. Personally, I like to separate professional matters from my private life at work
10. Respect and fulfilling my work on time and as required; I try not to get personally involved
G. I like departmental meetings because they create opportunities to build better relationships with colleagues and coworkers
15. Now I only establish work relationships during working hours; my free time belongs to my family
The perspective on employability in terms of interpersonal relationships suggests a preference for clearly separating work relationships from personal ones. In the workplace, they consider it essential to establish relationships that contribute to achieving professional goals

Source Own elaboration

can influence the candidate's perception of the organization and their willingness to accept a job offer. The literature suggests that negative feedback may be received with resistance and defensiveness if it is not delivered constructively and respectfully [19]. Conversely, positive feedback, such as praise regarding the candidate's skills and achievements, can enhance self-esteem and satisfaction with the interview [20]. Moreover, performance-related feedback during the interview can provide candidates with valuable information about their strengths and areas for improvement, which may shape their perception of organizational fit and their interest in the position [21].

Confidence when presenting the curriculum vitae during job interviews can significantly affect the overall impression recruiters have of a candidate and, therefore, their chances of success in the selection process. For this reason, *Question 16* (Table 4.16) *addresses this skill.* In the context of presenting the curriculum vitae during an interview, self-confidence may manifest through clear communication, assertive posture, and a confident demeanor [22]. Research has demonstrated that candidates who exhibit confidence when presenting their curriculum vitae are more likely to be perceived as competent and suitable for the position [23]. Conversely, expressing doubt during this process may generate distrust in the interviewer, diminishing interest in the applicant.

The use of tools to enhance labor market insertion opportunities is reflected in *Question 17* (Table 4.17), which aims to determine the extent to which respondents are willing to seek their own opportunities for improvement. Activities such as networking or engaging recruitment agencies are effective strategies to access hidden labor market opportunities and obtain privileged information, as highlighted by Higgins and Kram [24] in their research on the reconceptualization of career mentoring.

In addition to active participation in networking, job fairs, and professional events, individuals may increase their visibility by building a personal brand, which involves promoting their achievements, skills, and unique values, thereby enhancing their

Table 4.15 Question 15, SS questionnaire, labor market insertion

15. When you receive feedback during the interview, how do you take it?
5. When I have received feedback in interviews, it has generally been negative observations, which indicates to me that I will not be hired there
C. Through therapy haha—it was very difficult to understand at that time that the more experienced interviewers were pointing out my areas for improvement. I perceived it as criticism, and it bothered me
B. It was complicated, but I managed to balance the criticism and not take it into account if it was not relevant to me. I think I simply forgot about the companies that did not hire me
J. I still do not accept it easily, but I try to thank them when they provide it
7. Hahaha, I just thank them and take note of how I could improve
Regarding the acceptance of feedback in labor market insertion, it is observed that this skill develops over time and through experience. Most respondents indicated that initially they were bothered by feedback, but eventually they came to understand it as an opportunity for improvement, which allowed them to accept it without resentment

Source Own elaboration

Table 4.16 Question 16, SS questionnaire, labor market insertion

16. Do you feel confident when presenting your curriculum vitae in job interviews?
1. The first time, not confident at all, as I believed I lacked sufficient background to work as an engineer. However, over time my confidence in my abilities has grown
E. Not always. When I step outside of my comfort zone, I do not feel entirely confident. However, when asked about projects I had completed, I was able to mention some carried out at the institute
11. Yes, I trust in my abilities
Yes, especially when I conduct interviews in English, as I am fluent in it
O. My curriculum vitae was prepared by professionals on digital platforms. I simply filled the sections with my skills, competencies, and courses as required. Therefore, I trust it
is well-structured. Additionally, when I know it is a highly competitive company, I ask a friend to conduct a mock interview with me in advance, to anticipate potential questions
With respect to written communication in labor market insertion, respondents indicated that their experience in projects enables them to effectively present this in their curriculum vitae. Nevertheless, they emphasized that seeking assistance from various employment agencies helps improve their CVs depending on the position for which they are applying

Source Own elaboration

Table 4.17 Question 17, SS questionnaire, labor market insertion

17. Have you participated in networking activities or contacted recruiters to improve your opportunities for labor market insertion?
12. No, almost always it has been through recommendations from former classmates or colleagues
H. Only the first time I looked for a job, I used them to improve my CV
8. Only when I have sufficient time before actively seeking a job
15. No, never
10. Yes, having a company manages my curriculum and searching for new opportunities to grow and earn more is extremely important for my personal development
In terms of using networks or new technologies for labor market insertion, respondents reported relying on employment agencies to identify job opportunities. They also emphasized the usefulness of word-of-mouth communication and the importance of having recommendations from individuals within the organization itself

Source Own elaboration

attractiveness to employers [25]. Another approach, according to Yate [26], involves tailoring the curriculum vitae and cover letter for each specific position of interest.

For this reason, *Question 18* (Table 4.18) asked respondents whether they employ such tools to promote themselves within organizations.

Wanberg and Banas' [27] theory of work adjustment suggests that self-assessment of adaptability skills and the willingness to learn and grow in response to change are key elements to remain competitive in a dynamic labor market. Honest self-evaluation of strengths, weaknesses, and areas for improvement enables individuals to identify development opportunities and prepare to face new professional challenges [16]. For this reason, *Question 19* (Table 4.19) *examined their capacity for adaptation.* Those

Table 4.18 Question 18, SS questionnaire, labor market insertion

18. What measures do you take to improve your visibility in the job market and stand out as a desirable candidate for employers?
D. I now emphasize the projects I have completed, the certifications I have earned, and everything that adds value to my professional career
1. I seek direct recommendations from my friends
F. I include the certifications or courses I have completed to improve my engineering practice
14. None. I only include in my curriculum what I have already done
4. I keep my curriculum updated and always place first what is most relevant to the position I am applying for. I also mention the skills I have acquired, such as team management, communication, conferences, and providing training to other engineers
After experiencing multiple interviews during the process of labor market insertion and professional development, respondents acknowledged the importance of highlighting certifications and projects in which they have participated to showcase their professional expertise. Furthermore, they noted that they adapt their curriculum vitae to align with the specific position they are applying for

Source Own elaboration

who are able to self-assess effectively and adjust to labor market changes are better positioned to identify and capitalize on emerging job opportunities.

Question 20 (Table 4.20) *focuses on job satisfaction related to remuneration.* The initial salary that an individual receives upon entering the labor market can significantly influence their career progression and long-term financial stability. According to Schultz's [28] human capital theory, labor income is conditioned by factors such as education level, skills, and work experience. Consequently, a low initial salary may indicate insufficient investment in human capital formation prior to entering the labor market, which could limit future earning opportunities [29]. Moreover, initial wage disparities between different demographic and socioeconomic groups may perpetuate economic inequalities throughout professional life [30]. While the first salary is relevant, it is important to acknowledge that salary determination is

Table 4.19 Question 19, SS questionnaire, labor market insertion

19. How would you evaluate your capacity to adapt to changes in the labor market and take advantage of available employment opportunities?
14. My adaptability now depends more on patience. I pay closer attention to what companies offer and the opportunities for growth they provide
A. Quite well, since I dedicate myself to mastering daily tasks
G. I adapt with some difficulty, but after a few months, once I have proven my abilities, everything becomes easier to handle
11. As I am always undergoing training and seeking new opportunities, I believe I am very informed about the labor market
10. My adaptability has increased significantly because I do not work alone; I rely on employment search platforms
Responses regarding adaptation to change in the context of labor market insertion indicated that, although some experience difficulties in accepting them, teamwork enables more efficient and faster adjustment

Source Own elaboration

influenced by multiple factors, and that subsequent educational and career decisions can affect an individual's economic and professional mobility.

When entering the labor market, initial compensation is not necessarily the highest or most favorable. However, over time, respondents developed the ability to highlight their skills more effectively. After approximately one year, they tend to seek new employment opportunities that provide higher remuneration.

Once the specific responses and the corresponding general perspective were collected, the research problem, research question, objectives, and hypothesis established in the first chapter were addressed. This first chapter states that, since Electromechanical Engineering (IELMEC) predominantly focuses on the development of technical competencies while often neglecting aspects related to social skills, it is of vital importance to examine whether the soft skills acquired by students provide them with an advantage when entering the labor market within their specialization. The purpose of this study was to analyze the relationship between the soft skills developed by Electromechanical Engineering students at the Technological Institute of Ciudad Juárez and their usefulness for entering the labor market.

In response to the research question—*What is the relationship between soft skills as useful tools in the labor insertion process of Electromechanical Engineering students at the Technological Institute of Ciudad Juárez?*—within the specific context of the IELMEC program, which has traditionally emphasized technical skill development, it is essential to assess how the strengthening of soft skills influences students' ability to secure employment and perform successfully in their professional field. Employers value not only candidates with solid technical knowledge but also those with interpersonal and managerial skills that allow them to collaborate effectively in multidisciplinary teams, communicate efficiently with clients and colleagues, address problems creatively, and adapt to constantly changing work environments.

Therefore, considering the responses provided by the interviewees, it can be inferred that soft skills are closely related to labor insertion, as companies require

Table 4.20 Question 20, SS questionnaire, labor market insertion

20. Do you believe the salary you received was the best possible?
2. Yes, I believe it was fair for what I offered. It was above what the technicians earned, but I eventually had to change jobs to earn more money
K. No, but I accepted it while continuing to search until I found a position with the salary I wanted
9. Salary, salary… I think it is never enough. I always believe I can earn more
D. Yes, in fact, I earn quite well
3. Yes. Since the moment I receive interview invitations, I already have an idea of how much the company pays for the position I am applying for. I usually add 5 to 10% and start negotiations from there. It does not always work, but when it does, I win!
When entering the labor market, initial compensation is not necessarily the highest or most favorable. However, over time, respondents developed the ability to highlight their skills more effectively. After approximately one year, they tend to seek new employment opportunities that provide higher remuneration

Source Own elaboration

not only the hard skills acquired through academic training but also the human and social knowledge that empower engineers to emerge as leaders of their teams.

The proposed hypothesis suggests that the development of soft skills (SS) among students leads to valuable tools that generate employer interest, thereby facilitating their labor insertion into the local industry. This premise implies that soft skills—such as effective communication, teamwork, adaptability, and other interpersonal competencies—play a fundamental role in the hiring process and in shaping employers' perceptions of candidates.

Employers, when seeking candidates to fill job positions, value not only technical and academic competencies but also the soft skills possessed by applicants. It is therefore assumed that students who demonstrate mastery of these skills will have a competitive advantage in seeking employment in the local industry. This advantage may manifest as a greater predisposition among employers to consider these candidates during the selection process and, ultimately, to offer them employment opportunities.

Furthermore, it is presumed that the recognition and appreciation of soft skills by employers contributes to improving students' adaptability to the workplace and enhancing their long-term employability. Employers may perceive individuals with strong soft skills as assets capable of integrating quickly into work teams, collaborating effectively with colleagues and clients, and facing the changing challenges of the labor market with confidence and resilience. In this sense, the hypothesis suggests that the development of soft skills is essential not only for initial labor insertion but also for long-term professional growth and sustainability.

The objective was to investigate the relationship between soft skills and their usefulness as tools for students' labor insertion. The study sought to understand how soft skills such as effective communication, teamwork, adaptability, and other interpersonal competencies influence students' ability to secure employment in the labor market. This research focused on analyzing how the development and demonstration of these skills affect students' employment opportunities and their success in transitioning into the labor market.

The relationship between soft skills and students' labor insertion is a highly relevant topic in both educational and professional contexts. Soft skills, also referred to as interpersonal or socio-emotional skills, go beyond technical knowledge and are related to the way individuals interact, communicate, and work collaboratively.

At present, employers increasingly value soft skills when hiring new employees. This is because such skills are essential for success in the workplace, as they influence adaptability, problem-solving capacity, effective communication, and the ability to work as part of a team.

Students with well-developed soft skills have a higher probability of success in their labor insertion. For example, those capable of effective communication, stress management, conflict resolution, and teamwork are more highly valued by employers, as these skills are indispensable in any work environment.

Moreover, soft skills can also help students stand out during the job search process, as they highlight their capacity to adapt, learn, and collaborate in a professional setting. Therefore, developing and demonstrating soft skills can significantly increase students' labor insertion opportunities and enhance their long-term professional success.

The interview responses are connected as follows:

Soft skills are fundamental to labor insertion and professional success, as they influence various aspects of work life. The relationship between the soft skills questions (1–10) and the labor insertion questions (11–20) is established below.

The ability to communicate effectively with colleagues in the workplace (Question 1) relates to the perception of the importance of communication in labor insertion (Question 11), as both emphasize the relevance of communication skills in the professional environment.

Conflict management (Question 2) is linked to the acceptance of feedback in labor insertion (Question 15), as both involve the ability to resolve conflicts constructively and handle criticism productively.

Adaptation to workplace changes (Question 3) is related to adaptation to new work environments in general (Question 13), as both involve the ability to adjust and become familiar with new professional situations.

Teamwork (Question 4) is linked to perspectives on labor insertion regarding interpersonal relationships (Question 14), as both highlight the importance of building effective relationships and collaborating with others in the workplace.

Stress and pressure management at work (Question 5) relates to adaptation to changes in the context of labor insertion (Question 19), as both involve the ability to handle stressful situations and adapt to new professional challenges.

The ability to receive feedback (Question 6) is linked to how individuals adapt to new work environments in general (Question 13), as both involve the ability to learn and grow from experiences and feedback.

Self-motivation (Question 7) relates to adaptation to changes in the context of labor insertion (Question 19), as both involve the ability to remain motivated and persevere in the face of challenges and workplace changes.

Autonomy in decision-making (Question 8) is linked to the ability to adapt to changes in the context of labor insertion (Question 19), as both involve making independent decisions and adjusting to new professional circumstances.

Professional updating (Question 9) relates to written communication in labor insertion (Question 16), as both involve the ability to effectively communicate relevant experiences and skills in a résumé.

Self-awareness (Question 10) is linked to the perception of the importance of communication in labor insertion (Question 11), as both involve the ability to understand oneself and communicate personal skills and strengths effectively in the workplace.

Priorities in labor insertion, such as the importance of highlighting certifications and projects during the process (Question 18), reflect the consideration of aspects such as salary (Question 20) and company prestige before evaluating whether a position aligns with one's professional profile (Question 2).

Although all the questions related to soft skills (SS) could be linked to those concerning employability, it is considered important that they remain aligned within the same intention. The qualitative analysis was conducted narratively through the individual experiences of the participants, even though their responses occasionally coincided.

The analysis presented demonstrates both the prominence and the relative deficiencies in the development of soft skills (SS) and their relationship with employability among Electromechanical Engineering (IELMEC) students at the Technological Institute of Ciudad Juárez (ITCJ). This graphical representation provides a quantitative perspective on the competencies in which students have achieved a higher level of mastery and those in which their development is more limited. Such a diagnosis is fundamental to identifying areas of improvement and opportunities for personal and professional growth among future engineers. In particular, soft skills that exhibit incipient development require attention and specific strategies to foster their strengthening.

To address these deficiencies, it is essential to implement programs and workshops focused on the development of skills such as effective communication, teamwork, and conflict resolution. These aspects are crucial not only for academic training but also for the successful employability of students in an increasingly competitive and demanding labor market. By promoting practical and collaborative learning, students can be better prepared to face the challenges of the professional world with confidence and efficacy. Furthermore, it is important for both educational institutions and companies to collaborate in creating practice and development opportunities, where students can apply and refine their soft skills in real-world situations. This would not only enrich their formative experience but also increase their employability and adaptability to the workplace. In this way, comprehensive training can be achieved, encompassing both technical competencies and interpersonal skills, ensuring that graduates are fully prepared for the challenges they will face in their professional careers.

Two noteworthy aspects in the training of Electromechanical Engineering (IELMEC) students are Continuous Learning and Teamwork combined with Emotional Intelligence. These competencies have proven fundamental for their successful integration into the labor market. According to interviews conducted, participation in student projects has been crucial for the development of these skills. At the beginning of their studies, many students faced difficulties relating to others and managing differences in personality. Both teamwork and emotional intelligence require autonomous decision-making capacity, as well as conflict resolution skills, project leadership, effective communication, and the motivation to achieve goals.

These skills are highly valued by employers, as they represent essential requirements in the workplace. The ability of IELMEC students to perform both autonomously and collaboratively is especially relevant, considering that many local companies expect their personnel to possess such competencies from the outset. Figure. 4.1 presents an analysis of the level of development of these skills.

Among the soft skills in which deficiencies were identified—representing areas for growth in students' personal development—are adaptability, autonomy, receptivity to feedback, and time and stress management. These skills are also highly valued by employers, who acknowledge that their full development often occurs in professional contexts rather than during academic studies. Although students demonstrate

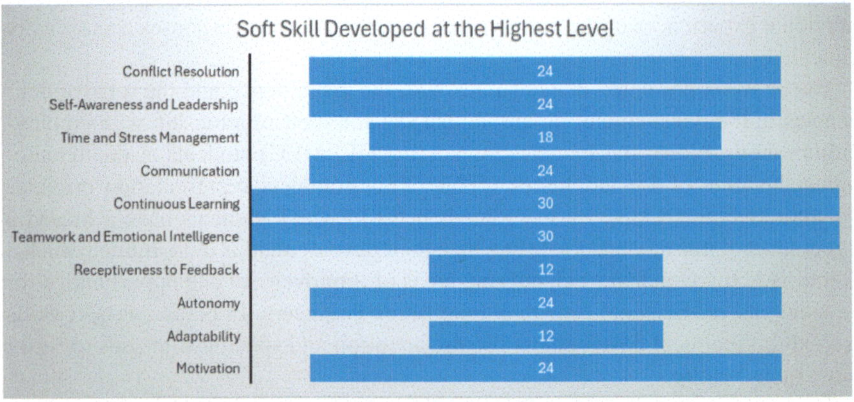

Fig. 4.1 Soft skills developed at the highest level. Own elaboration

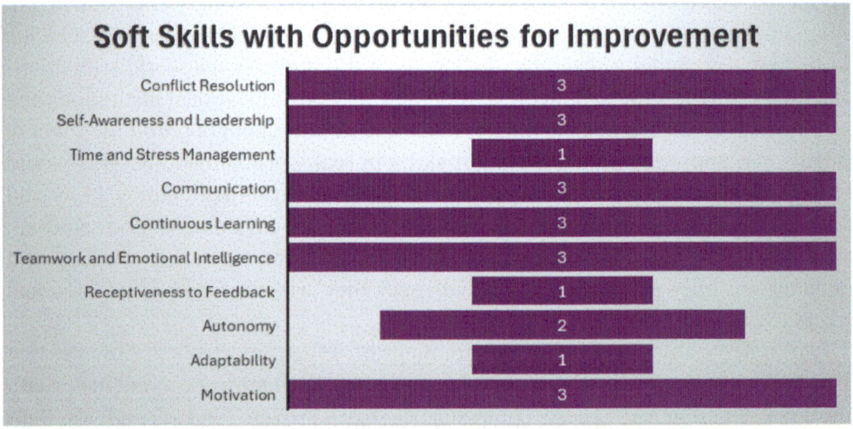

Fig. 4.2 Soft skills with opportunities for improvement. Own elaboration

the ability to manage stress and time when completing projects within the deadlines set by professors, in the industry these skills acquire greater significance and urgency due to the critical nature of many proprietary projects. Professional experience provides a more demanding environment for the development of these skills. Figure 4.2 illustrates the areas where deficiencies in these skills have been identified as opportunities for improvement.

References

1. Robbins, S.P., Judge, T.A.: Organizational Behavior, 17th edn. Pearson Education, USA (2019)
2. Ivancevich, J.M., Konopaske, R.: Organizational Behavior and Management. Mexico. McGraw-Hill Interamericana (2013)
3. Stone, D., Patton, B., Heen, S.: Difficult Conversations: how to Discuss What Matters Most. Penguin Books (2010)
4. Deci, E.L., Ryan, R.M.: Intrinsic Motivation and Self-determination in Human Behavior. Plenum, New York (1985)
5. Locke, E.A., Latham, G.P.: A Theory of Goal Setting & Task Performance. Prentice Hall, Englewood Cliffs, NJ (1990)
6. Kirkpatrick, S.A., Locke, E.A.: Direct and indirect effects of three core charismatic leadership components on performance and attitudes. J. Appl. Psychol. **81**(1), 36–51 (1996)
7. Noe, R.A.: Employee Training and Development. McGraw-Hill/Irwin (2010)
8. Durlak, J.A., Weissberg, R.P., Dymnicki, A.B., Taylor, R.D., Schellinger, K.B.: The impact of enhancing students' social and emotional learning: a meta-analysis of school-based universal interventions. Child Dev. **82**(1), 405–432 (2011)
9. Heckman, J.J., Stixrud, J., Urzua, S.: The effects of cognitive and noncognitive abilities on labor market outcomes and social behavior. J. Law Econ. **24**(3), 411–482 (2006)
10. Dionigi, R.A., Littlejohn, A.: Placing the self in professional communication. J. Bus. Tech. Commun. **22**(4), 399–422 (2008)
11. Deci, E.L., Ryan, R.M.: The "what" and "why" of goal pursuits: human needs and the self-determination of behavior. Psychol. Inq. **11**(4), 227–268 (2000)
12. Kristof-Brown, A.L., Zimmerman, R.D., Johnson, E.C.: Consequences of individuals' fit at work: a meta-analysis of person-job, person-organization, person-group, and person-supervisor fit. Pers. Psychol. **58**(2), 281–342 (2005)
13. Edwards, J.R., Cable, D.M.: The value of value congruence. J. Appl. Psychol. **94**(3), 654–677 (2009)
14. Bauer, T.N., Bodner, T., Erdogan, B., Truxillo, D.M., Tucker, J.S.: Newcomer adjustment during organizational socialization: a meta-analytic review of antecedents, outcomes, and methods. J. Appl. Psychol. **92**(3), 707–721 (2007)
15. Ashford, S.J., Rothbard, N.P., Piderit, S.K., Dutton, J.E.: Out on a limb: the role of context and impression management in selling gender-equity issues. Adm. Sci. Q. **48**(3), 485–516 (2003)
16. Tims, M., Bakker, A.B., Derks, D.: Daily job crafting and the self-efficacy-performance relationship. J. Manag. Psychol. **31**(1), 5–21 (2016)
17. Erdogan, B., Bauer, T.N.: Differentiated leader-member exchanges: the buffering role of justice climate. J. Appl. Psychol. **95**(6), 1104–1120 (2010)
18. Wayne, S.J., Liden, R.C., Kraimer, M.L., Graf, I.K.: The role of human capital, motivation and supervisor sponsorship in predicting career success. J. Organ. Behav. **23**(2), 247–266 (2002)
19. Tedeschi, J.T., Melburg, V.: Impression management and influence in the organization. Res. Organ. Behav. **6**, 123–171 (1984)
20. Ilgen, D.R., Fisher, C.D., Taylor, M.S.: Consequences of individual feedback on behavior in organizations. J. Appl. Psychol. **64**(4), 349–371 (1979)
21. Latham, G.P., Brown, T.C.: The effect of learning vs. outcome goals on self-efficacy, satisfaction and performance in an MBA program. Appl. Psychol. **91**(4), 840–849 (2006)
22. Judge, T.A., Hurst, C.: Capitalizing on one's advantages: role of core self-evaluations. J. Appl. Psychol. **92**(5), 1212–1227 (2007)
23. Anderson, N., Shackleton, V.J.: Selection and performance: the impact of ability, personality, and the situation. J. Occup. Organ. Psychol. **66**(3), 219–243 (1993)
24. Higgins, M.C., Kram, K.E.: Reconceptualizing mentoring at work: a developmental network perspective. Acad. Manag. Rev. **26**(2), 264–288 (2001)
25. Peters, T.J.: The brand called you. Fast Company **10**, 83–94 (1997)
26. Yate, M.: Knock 'em Dead Resumes: a killer resume gets more job interviews! Adams Media (2017)

27. Wanberg, C.R., Banas, J.T.: Predictors and outcomes of openness to changes in a reorganizing workplace. J. Appl. Psychol. **85**(1), 132–142 (2000)
28. Schultz, T.W.: Investment in human capital. Am. Econ. Rev. **51**(1), 1–17 (1961)
29. Mincer, J.: Schooling, Experience, and Earnings. Columbia University Press, New York (1974)
30. Blau, F.D., Kahn, L.M.: The gender wage gap: extent, trends, and explanations. J. Econ. Lit. **55**(3), 789–865 (2017)

Chapter 5
Conclusions Part I

The aspects discussed address a series of crucial considerations related to labor market integration and the cultivation of soft skills within this context. First, emphasis is placed on the relevance of effective communication, which is recognized as a key element when interacting with potential employers during the job search process. This perspective aligns with the notion that communication skills are essential and have been developed for application in professional environments.

Companies demonstrate a tendency to allocate both temporal and financial resources toward strengthening the skills of their workforce. However, there is the possibility that these entities may adopt optimization strategies when identifying individuals whose soft skills (SS) are already fully developed, allowing them to achieve significant savings. Nonetheless, the lack of specific professional competencies may motivate organizations to consider the selection of candidates who demonstrate these skills as a complement. Therefore, the process of labor market incorporation is influenced by the interactive dynamics between two key entities: the company and the candidate. Personality traits displayed during the interview procedure play a decisive role in the final hiring evaluation.

Additionally, candidates applying for employment positions acknowledge the consideration of a diverse range of factors during the selection process, such as the salary offered, as well as the reputation and size of the company. These elements are assessed to determine suitability and compatibility with professional profile. This approach suggests that individuals consider a variety of criteria when seeking employment, which transcend mere alignment with academic training.

The process of labor market integration involves a series of stages and dynamics that influence the interaction between candidates and companies. Organizations are observed to show a predisposition to invest resources in the development of their personnel's skills, although they may opt to economize if they identify individuals whose soft skills are already developed. Nonetheless, the absence of specific professional skills may lead to the consideration of candidates who possess these competencies as a complement. In this regard, the process of labor market integration is

G. H. Esquivel and P. A. Hernández, *Bridging the Gap: Comprehensive Project Design for Soft Skills Development and Labor Market Integration*, Information Systems Engineering and Management 82, https://doi.org/10.1007/978-3-032-19026-0_5

shaped by the interaction between two key agents: the company and the candidate, with personality dynamics serving as a crucial factor in the final hiring assessment.

Moreover, job candidates recognize the importance of diverse factors during the selection process, such as salary levels, company reputation, and organizational size, in determining compatibility with their professional profile. This approach reflects the consideration of multiple criteria when seeking employment, extending beyond academic qualifications. Furthermore, regarding adaptation to new work environments, it is emphasized that this process tends to be brief, characterized by rapid familiarization with colleagues and the workplace setting. Although initial challenges may arise, these are often overcome more effectively through teamwork and collaboration.

The importance of receiving and accepting feedback in workplace environments is highlighted as a skill that evolves with experience and time. Although it may initially present difficulties, as individuals mature personally, they learn to perceive feedback as an opportunity for growth and continuous improvement.

The significance of this process in the professional sphere is recognized, where effective feedback contributes to development and work efficiency. The ability to receive constructive criticism and assimilate it as a path for progress constitutes a valuable competency that is nurtured through practice and reflection.

The complexity and diversity of soft skills that influence labor market integration are underscored as a fundamental aspect of professional insertion. From effective communication to flexibility in the face of change and the capacity to accept constructive criticism, these competencies form a crucial framework during the job search and adaptation to workplace environments. Their interaction and development exert a significant influence on individuals' performance and professional progress, highlighting the importance of their mastery and continuous improvement.

In this sense, it is evident that the transition to the labor market is closely linked to the ability to manage a variety of soft skills. These skills, ranging from effective communication to adaptability and resilience in the face of feedback, intertwine in complex ways throughout the job search and integration into the workplace. The development and effective application of these competencies play a fundamental role in individuals' success and professional growth, underscoring the importance of their comprehension and ongoing refinement.

Part II
Comprehensive Project Design for the Development of Soft Skills

This study explores the relationship between soft skills (SS) and labor market integration (LMI) among students of the Electromechanical Engineering program (IELMEC) at the Technological Institute of Ciudad Juárez (ITCJ). Evidence shows that many professionals work outside their field of specialization due to limited opportunities and insufficient career planning, a situation confirmed by national statistics. Employers demand both technical skills (TS) and soft skills (SS), yet the latter are not systematically developed within engineering programs, creating a gap between graduate profiles and industry requirements.

The research is structured in chapters that address the problem statement, research questions, objectives, justification, and hypotheses (Chap. 6), followed by the theoretical framework (Chap. 7). The methodology is presented in Chap. 8, including research design, diagnostic instruments, and statistical analysis of soft skills. Chapter 9 evaluates the outcomes of project-based learning in statics through a soft skills progress questionnaire and results analysis. Finally, Chap. 10 outlines conclusions and recommendations to improve professional training and employability.

Findings highlight the importance of integrating soft skills such as communication, adaptability, leadership, and problem-solving into engineering education. Active methodologies like challenge-based learning (CBL) foster the connection between theory and practice, as supported by authors including Haro, Ayala, Núñez, and Sital. Moreover, international reports, such as those by the World Economic Forum, affirm that these skills will remain essential in future labor markets. Consequently, the implementation of integrative projects enhances graduates' employability by combining technical expertise with human development.

Chapter 6
Introduction

In the current context of higher education, soft skills (SS) have become an essential component for the insertion and professional success of graduates. In a recent study conducted at the Technological Institute of Ciudad Juárez (ITCJ), five key soft skills were identified as fundamental by employers in the field of Electromechanical Engineering: effective communication, teamwork, leadership, problem-solving, and adaptability (Ayala-Hernández and Haro-Esquivel 2024).

The results of this research show that these skills are not systematically developed during academic training. In response to this need, the proposal emerges to design integrative projects that, when implemented within the subjects of the curriculum, promote the intentional strengthening of these skills. This strategy seeks for each project to focus specifically on the development of one or more of these competencies, aligning academic objectives with the real demands of the labor environment.

6.1 Problem Statement

Despite advances in the technical training of Electromechanical Engineering students, a significant gap persists between the soft skills (SS) demanded by employers and those developed by students during their academic trajectory. Previous research conducted at the Technological Institute of Ciudad Juárez (ITCJ) has identified a specific set of priority soft skills, such as effective communication, teamwork, problem-solving, adaptability, and leadership (Ayala-Hernández and Haro-Esquivel 2024).

However, current curricula and pedagogical practices do not comprehensively or systematically include the design of projects aimed at the intentional strengthening of these competencies. This lack of alignment limits the labor insertion and professional performance of graduates in increasingly demanding and collaborative industrial

G. H. Esquivel and P. A. Hernández, *Bridging the Gap: Comprehensive Project Design for Soft Skills Development and Labor Market Integration*, Information Systems Engineering and Management 82, https://doi.org/10.1007/978-3-032-19026-0_6

environments. Therefore, it is necessary to design and implement integrative projects that promote the structured development of these skills within the academic context.

6.2 Research Questions

1. What specific soft skills, previously identified, show the greatest deficit among Electromechanical Engineering students at ITCJ?
2. What characteristics should integrative projects have to effectively foster the development of these soft skills?
3. What impact does the implementation of integrative projects have on strengthening the soft skills demanded by the employer sector?

6.3 General Objective

To design and implement integrative projects that foster the development of soft skills in Electromechanical Engineering students at ITCJ, based on the demands identified by the employer sector.

6.4 Specific Objectives

- Identify soft skills with the greatest need for strengthening among students.
- Establish guidelines and methodologies for the design of integrative projects aimed at SS development.
- Evaluate the impact of integrative projects on the development of these skills.

6.5 Justification

The labor insertion of Engineering graduates does not depend exclusively on their technical knowledge but also on their ability to interact, adapt, lead, and solve problems in real environments. Soft skills are currently recognized as key competencies by international organizations [1], employers [2], and educational institutions. This study will contribute to the design of pedagogical strategies applicable in the classroom, aligned with labor market demand, generating more competitive graduates prepared for the challenges of the twenty-first century.

6.6 Hypotheses

Alternative Hypothesis (HA): The implementation of integrative projects designed based on the soft skills requested by employers significantly improves the development of these skills in Electromechanical Engineering students at ITCJ.

Null Hypothesis (H0): The implementation of integrative projects does not produce significant improvements in the development of soft skills in Electromechanical Engineering students.

References

1. OECD.: The future of education and skills 2030 (2020). https://www.oecd.org/education/2030-project/
2. Robles, M.M.: Executive perceptions of the top 10 soft skills needed in today's workplace. Bus. Commun. Q. **75**(4), 453–465 (2012). https://doi.org/10.1177/1080569912460400

Chapter 7
Theoretical Framework

At present, the training of engineers requires not only technical and scientific mastery but also the development of soft skills that enable them to integrate effectively into multidisciplinary teams, solve complex problems, and communicate clearly in industrial contexts. Soft skills, understood as interpersonal, communicative, organizational, and adaptive capacities, have become key competencies for employability and professional performance [1, 2].

Authors such as Haro Esquivel and Ayala Hernández [3], in their work *Mechatronic Engineering: Soft Skills for Workforce Integration* highlight that successful insertion into the labor market of the border region, such as Ciudad Juárez, largely depends on competencies such as effective communication, leadership, adaptability, collaboration, and the ability to solve problems in high-pressure environments. These authors emphasize that engineers must be trained with active methodologies that simulate real industrial scenarios. From the educational perspective, it is noted that integrative projects aligned with the competency-based approach are effective for linking theory with professional practice, as they allow students to develop concrete products, solve real problems, and work in multidisciplinary teams. This author proposes formative assessment as an essential strategy to measure progress in the development of soft skills.

In turn, it is argued that challenge-based learning (CBL) is particularly valuable in the context of engineering, as it promotes critical reflection, structured thinking, and effective communication. CBL encourages students to assume shared responsibilities and make well-founded technical decisions.

At the international level, studies such as those of Succi and Canovi [2] confirm that companies increasingly value transversal skills in engineering graduates, even above certain specific technical knowledge. Likewise, the World Economic Forum [4] indicates that by 2025, soft skills such as problem-solving, communication, teamwork, and emotional intelligence will rank among the most desired competencies by employers.

G. H. Esquivel and P. A. Hernández, *Bridging the Gap: Comprehensive Project Design for Soft Skills Development and Labor Market Integration*, Information Systems Engineering and Management 82, https://doi.org/10.1007/978-3-032-19026-0_7

In the context of globalization and the growing complexity of the industrial environment, the training of engineers requires complementing technical mastery with soft skills or transversal competencies. Contemporary evidence indicates that the absence of these skills significantly limits the insertion and professional success of engineers in today's labor market.

These skills encompass abilities such as effective communication, leadership, adaptability, time management, teamwork, emotional intelligence, and creative problem-solving. Various studies have confirmed that companies assign equal or greater importance to these competencies compared to traditionally valued technical skills. The development of soft skills within the field of engineering is essential to close the gap between university and industry, as these capacities are crucial to facing the challenges of the modern labor market.

In this context, the implementation of integrative projects in the training of electromechanical engineers enables the articulation of technical knowledge with the development of soft skills through active methodology, linkage with the productive sector, and constant feedback. These strategies enhance the academic and professional relevance of graduates in work environments such as Ciudad Juárez, where the manufacturing industry demands highly adaptable, technically competent, and socially collaborative profiles.

The development of soft skills in the educational field has become a strategic priority for educational systems that aim to train competent, adaptable individuals capable of facing the challenges of the twenty-first century. These skills, also known as socioemotional competencies, include aspects such as communication, leadership, collaboration, problem-solving, self-regulation, and empathy [5]. Unlike technical skills, soft skills are transferable to any occupational field and are considered fundamental for employability and civic life.

According to Barni de Campos [6], recent studies highlight that at least 85% of the most desired skills for employability in engineering correspond to soft skills, while only 15% are technical. International organizations such as the World Economic Forum emphasize that by 2025, competencies such as collaborative work, emotional intelligence, critical thinking, and adaptability will occupy the top positions in professional requirements lists.

In this context, Project-Based Learning (PBL) emerges as an effective pedagogical methodology to foster the development of such skills. This approach is based on the idea that students learn more meaningfully when they engage in projects that solve real or authentic problems, allowing them to integrate theoretical and practical knowledge in collaborative situations [7]. PBL promotes student-centered teaching, where the teacher's role shifts to that of a facilitator of learning, guiding the process through questions, challenges, and constant feedback.

Various studies have demonstrated that PBL contributes to the strengthening of soft skills. For example, Barron and Darling-Hammond [8] point out that, when working on projects, students develop competencies such as critical thinking, oral and written communication, teamwork, and time management. These skills are not only relevant in the educational context but are also highly valued in the professional field.

Likewise, Bell [9] highlights that projects allow students to assume responsibilities within a group, participate in decision-making, and resolve conflicts, which strengthens their capacity for leadership, empathy, and reflective thinking. This process does not occur spontaneously but must be planned with pedagogical intentionality, ensuring that the projects address real challenges and that their development includes moments of reflection and evaluation.

From a constructivist perspective, PBL is grounded in the idea that knowledge is actively and socially constructed. According to Blumenfeld et al. [10], participation in collaborative tasks enhances students' intrinsic motivation and promotes deeper learning. Furthermore, interdisciplinary work allows students to face complex situations from diverse perspectives, strengthening their systemic thinking and capacity to work with others.

Assessment in the context of PBL also plays a crucial role. According to Thomas [7], it should encompass both the final product of the project and the work process, valuing individual and group performance. Continuous feedback allows students to become aware of their progress, identify areas for improvement, and strengthen their self- knowledge, which is key to the development of metacognitive and socioemotional skills.

In summary, Project-Based Learning offers an ideal pedagogical environment for the development of soft skills, by fostering active participation, collaboration, and reflection. Its implementation, however, requires rigorous planning, adequate teacher training, and a school culture that values both cognitive and socioemotional learning. This methodology represents a transformative strategy for training competent professionals and critical citizens capable of facing the challenges of a constantly changing world.

Furthermore, the integration of real-world engineering scenarios within Project-Based Learning enhances its effectiveness by providing students with authentic contexts in which to apply theoretical knowledge. By engaging with practical problems, learners not only reinforce technical competencies but also develop adaptability, problem-solving, and communication skills in a manner that mirrors professional environments. This experiential approach ensures that graduates are better prepared for the demands of the labor market, fostering both technical proficiency and the interpersonal abilities essential for successful multidisciplinary collaboration [1, 3].

References

1. Robles, M.M.: Executive perceptions of the top 10 soft skills needed in today's workplace. Bus. Commun. Q. **75**(4), 453–465 (2012). https://doi.org/10.1177/1080569912460400
2. Succi, C., Canovi, M.: Soft skills to enhance graduate employability: comparing students' and employers' perceptions. Stud. High. Educ. **45**(9), 1834–1847 (2020). https://doi.org/10.1080/03075079.2019.1585420
3. Haro Esquivel, G., Ayala Hernández, P.: Mechatronic Engineering: soft Skills for Labor Insertion. Spanish Academic Publishing (2024)

4. World Economic Forum.: The future of jobs report 2020 (2020). https://www.weforum.org/reports/the-future-of-jobs-report-2020
5. OECD.: Future of education and skills 2030: OECD learning compass 2030 (2019). https://www.oecd.org/education/2030
6. Barni de Campos, D., Mauricio Martins de Resende, L., Borges Fagundes, A.: The importance of soft skills for engineering. Creat. Educ. (2020). http://www.scirp.org/journal/ce
7. Thomas, J.W.: A review of research on project-based learning. The Autodesk Foundation (2000). https://www.pblworks.org/sites/default/files/litreview.pdf
8. Barron, B., Darling-Hammond, L.: Teaching for meaningful learning: a review of research on inquiry-based and cooperative learning. George Lucas Educational Foundation (2008). https://files.eric.ed.gov/fulltext/ED539399.pdf
9. Bell, S.: Project-based learning for the 21st century: skills for the future. Clear. House: J. Educ. Strateg. Issues Ideas **83**(2), 39–43 (2010). https://doi.org/10.1080/00098650903505415
10. Blumenfeld, P.C., Soloway, E., Marx, R.W., Krajcik, J.S., Guzdial, M., Palincsar, A.: Motivating project-based learning: sustaining the doing, supporting the learning. Educ. Psychol. **26**(3–4), 369–398 (1991). https://doi.org/10.1080/00461520.1991.9653139

Chapter 8
Methodology

This chapter presents the methodological design employed to carry out the diagnostic process with students of Electromechanical Engineering (IELMEC). The primary objective of this stage was to identify which soft skills require reinforcement or further development in order to align student competencies with the demands of the labor market. By applying validated diagnostic instruments, the research seeks to obtain reliable data that reflects both the current level of soft skills among students and the areas in which additional pedagogical strategies may be needed.

The methodology adopted ensures a systematic and rigorous approach, combining quantitative and qualitative techniques to provide a comprehensive understanding of the problem. This process not only establishes the foundation for analyzing students' profiles in relation to professional expectations but also supports the design of educational interventions aimed at strengthening essential skills such as communication, adaptability, leadership, teamwork, and problem-solving. In this way, the methodological framework serves as a bridge between theory and practice, guiding the study toward evidence-based conclusions and practical recommendations.

8.1 Type of Research

This study is applied in nature, with a quantitative-qualitative (mixed) approach. It employs a quasi-experimental design with a control group and an experimental group to evaluate the impact of integrative projects, complemented by semi-structured interviews to gather perceptions and experiences from students and instructors.

1. Population and sample: Electromechanical Engineering students at ITCJ. A representative sample of at least two academic groups will be selected.
2. Data collection techniques: Diagnostic surveys, checklists, interviews, and evaluation rubrics.

G. H. Esquivel and P. A. Hernández, *Bridging the Gap: Comprehensive Project Design for Soft Skills Development and Labor Market Integration*, Information Systems Engineering and Management 82, https://doi.org/10.1007/978-3-032-19026-0_8

3. Phases of the study:

   - Diagnosis of soft skills.
   - Design and validation of integrative projects.
   - Implementation during the semester.
   - Comparative pre- and post-evaluation.

8.2 Expected Results

It is expected that students participating in integrative projects will show significant improvements in their performance of soft skills, particularly in communication, leadership, and collaborative work. In addition, a replicable proposal for the design of integrative projects for other courses in the program will be generated.

8.3 Activity Schedule

The following table presents a structured action plan designed to guide the systematic implementation of the research project. It outlines the sequence of activities, their objectives, and the timeline for completion, ensuring that each stage contributes effectively to the development, assessment, and analysis of students' soft skills. This plan serves as a framework for organizing tasks in a coherent and methodical manner, supporting both the research process and the achievement of the study's objectives (see Table 8.1).

Table 8.1 Scheduled activities. Author's elaboration

Activity	Week 1	Week 2	Week 3	Week 4	Week 5	Week 6
Initial diagnosis	X					
Project design	X					
Classroom implementation	X	X				
Final evaluation		X	X	X		
Systematization of results				X	X	
Final report writing					X	X

8.4 Diagnostic Instrument

The Soft Skills Diagnostic Questionnaire (initial version), presented in Table 8.2, is designed to evaluate the development of key competencies among Electromechanical Engineering (IELMEC) students. The instrument focuses on five essential soft skills: effective communication, teamwork, problem-solving, adaptability, and leadership. By systematically assessing these areas, the questionnaire provides insight into students' current strengths and identifies skills that require further development. This diagnostic tool serves as a foundation for targeted educational strategies and interventions aimed at enhancing students' professional readiness and labor market integration.

This questionnaire will be administered both before and after the implementation of the integrative projects to evaluate students' progress in the development of soft skills over the course of the semester.

The diagnostic was conducted with 56 students enrolled in the Statics course during the second semester of the Electromechanical Engineering program. The responses obtained provided valuable data regarding the initial level of soft skills and the areas requiring further reinforcement.

8.5 Statistical Analysis by Soft Skill in the Diagnostic Questionnaire

Each soft skill is assessed through three specific items designed to capture different aspects of competency. To quantify students' performance, the overall average for each skill was calculated by summing the individual responses and dividing the total by the number of participants (n = 56), as presented in Table 8.3. This calculation provides a clear overview of the general proficiency level for each skill within the group, allowing for the identification of strengths and areas requiring further development. Such analysis serves as a basis for guiding educational strategies and interventions aimed at enhancing students' soft skills in alignment with labor market demands.

The analysis of the results presented in Table 8.3 reveals the relative development of soft skills among Electromechanical Engineering students, highlighting the areas that require prioritized attention. Leadership was identified as the lowest-rated skill (2.72), indicating an urgent need for interventions aimed at enhancing decision-making, initiative, and the capacity to guide and motivate others. Effective communication (2.88) also presents notable challenges, particularly in public speaking and the preparation of technical reports, suggesting the need for structured exercises to strengthen both oral and written communication competencies. Adaptability, with an average score of 3.10, reflects a moderate level of development, yet it points to the importance of fostering positive attitudes toward change and flexibility in professional contexts. Problem-solving (3.32) is considered acceptable but could be further

Table 8.2 Diagnostic questionnaire for soft skills. Author's elaboration

	5	4	3	2	1
Section 1: Effective communication					
1. I express myself clearly in oral presentations					
2. I write technical reports or formal emails in a clear and structured way					
3. I actively listen to my peers and respect their opinions					
Section 2: Teamwork					
4. I collaborate with my peers to achieve common goals					
5. I know how to assume different roles within a team					
6. I show willingness to support others in their tasks					
Section 3: Problem-solving					
7. I easily identify problems in technical or academic situations					
8. I propose viable and creative solutions to technical challenges					
9. I evaluate the outcomes of my decisions and adjust when necessary					
Section 4: Adaptability					
10. I adapt quickly to changes in work or school projects					
11. I maintain a positive attitude in the face of uncertainty					
12. I learn new tools or methods when required					
Section 5: Leadership					
13. I feel comfortable leading a work group					
14. I take responsibility for making decisions for the common good					
15. I motivate my peers to achieve the team's objectives					

Instructions: Mark the option that best describes your level of mastery or experience in each skill. Use the scale from 1 to 5, where:

1 = Never or Very Poorly Developed. 2 = Rarely Developed or Developed at a Low Level. 3 = Sometimes Developed or Moderately Developed. 4 = Frequently developed or well developed. 5 = Always or fully developed

Table 8.3 Statistical analysis by soft skill. Author's elaboration

Soft skill	Items evaluated	General average
Leadership	Items 13, 14, 15	2.72
Effective communication	Items 1, 2, 3	2.88
Adaptability	Items 10, 11, 12	3.10
Problem solving	Items 7, 8, 9	3.32
Teamwork	Items 4, 5, 6	3.46

reinforced through guided technical challenges that integrate theoretical knowledge with practical applications. Teamwork, while the strongest skill (3.46), still shows room for improvement through collaborative projects and peer-feedback activities.

This ordered evaluation of soft skills provides a clear roadmap for prioritizing educational strategies, ensuring that resources are focused on addressing the most critical deficits while further enhancing existing competencies.

The following Table 8.4 presents recommendations intended to guide the design of integrative projects aimed at enhancing the development of key soft skills among Electromechanical Engineering students. Each suggested strategy is aligned with a specific skill and provides practical approaches to strengthen students' competencies in real-world contexts. By incorporating these targeted activities into project design, educators can effectively foster leadership, effective communication, adaptability, problem-solving, and teamwork, ensuring that students not only acquire technical knowledge but also develop the interpersonal and cognitive abilities essential for successful integration into the labor market.

Table 8.4 Recommendations for the design of integrative projects. Author's elaboration

Focal skill	Suggested strategy
Leadership	Design projects where students rotate the role of technical leader or coordinator
Effective communication	Include oral presentations, interview simulations, and guided technical writing
Adaptability	Propose activities with uncertain variables and unexpected contextual changes
Problem solving	Use real cases of failures or mechanical problems for collaborative analysis
Teamwork	Encourage interdisciplinary teams with common goals and distributed roles

Complementary Statistical Notes

- Mean standard deviation: ±0.64 (indicates moderate variability in responses).
- Data skewness: Slight negative tendency in leadership and communication (more students rated themselves low).
- Observed correlation: Positive relationship between communication and leadership ($r = 0.68$).

Chapter 9
Design and Outcomes of Integral Project Based Learning in Statics

Building on the results obtained from the diagnostic instrument applied to Electromechanical Engineering students, a series of five short-duration integrative projects were designed, each with a maximum duration of four hours. These projects were specifically conceived to address the soft skills identified as requiring reinforcement, allowing students to develop competencies that complement their technical knowledge. Each project integrates theoretical and practical content from the Statics course with the intentional cultivation of a transversal skill, ensuring that learning is both relevant and applicable to real-world engineering contexts.

The design of these projects follows a competency-based approach, emphasizing active learning strategies that engage students in problem-solving, collaboration, and decision-making. By situating technical challenges within a framework that also promotes leadership, effective communication, adaptability, teamwork, and problem-solving, the projects aim to create meaningful learning experiences that enhance professional readiness. The outcomes of these interventions provide insight into how structured project-based learning can foster both technical proficiency and the development of essential soft skills, thereby supporting the holistic formation of future electromechanical engineers.

Project 1: "Leadership Simulation in Structural Assembly"

Focal skill: *Leadership.*

Estimated duration: 4 h.

Objective: For the student to assume technical leadership of a team to solve a structural challenge.

Activity: The class will be divided into teams of four members. Each team will be assigned the theoretical construction and simulation of an isostatic structure (using software such as ForceEffect or SolidWorks Simulation). The team leader must distribute roles, make technical decisions, and present the final result.

G. H. Esquivel and P. A. Hernández, *Bridging the Gap: Comprehensive Project Design for Soft Skills Development and Labor Market Integration*, Information Systems Engineering and Management 82, https://doi.org/10.1007/978-3-032-19026-0_9

Evidence: Technical report and oral presentation by the leader.

Justification: "Technical leadership must be exercised in simulated contexts that represent real situations so that the student learns to guide from knowledge" [1].

Project 2: "Present Your Mechanism"

Focal skill: *Effective communication.*

Estimated duration: 3.5 h.

Objective: To develop the ability to communicate technical concepts both orally and in writing.

Activity: Each student will individually design a simple mechanism in equilibrium (lever, crane, articulated arm), calculate the forces, and draft a brief report (one page). Subsequently, the student will explain its functioning in a three-minute presentation before the class using graphic resources.

Evidence: Written report and oral presentation.

Justification: "Effective communication enables the sharing of technical knowledge in a clear and precise manner" [2].

Project 3: "Solving Failures: A Statics Case"

Focal skill: Problem solving.

Estimated duration: 4 h.

Objective: To apply knowledge of statics to solve a complex technical problem.

Activity: Teams will analyze a structural failure (provided as a case study, e.g., a collapsed bridge, unbalanced crane, etc.). They will determine the causes from the perspective of equilibrium, moments, and loads, and propose solutions with justification.

Evidence: Free-body diagram, written analysis, and proposed solution.

Justification: "Solving contextualized problems fosters critical thinking and technical decision-making" (Ayala-Hernández and Haro-Esquivel 2024).

Project 4: "Engineering in Motion"

Focal skill: Adaptability.

Estimated duration: 3.5 h.

Objective: To develop the ability to adjust to changes during a technical challenge.

Activity: Teams will solve an equilibrium problem for a structure. During the activity, the instructor will modify one condition (e.g., material change, support point modification, etc.), requiring students to reformulate the solution.

Evidence: Logbook of adaptation to change and new analysis.

Justification: "Adaptability is strengthened when students are exposed to unforeseen scenarios within a safe framework" [3].

Project 5: "Collaborative Construction of a Tower"

Focal skill: *Teamwork.*

Estimated duration: 4 h.

Objective: To foster effective collaboration through a task with a technical purpose.

Activity: Teams will construct a tower using simple materials (pasta, sticks, string, tape) capable of supporting a given load. Teamwork, coordination, application of statics principles, and stability will be evaluated.

Evidence: Peer evaluation of group performance and load testing.

Justification: "Collaborative work enhances efficiency, innovation, and shared responsibility" [2].

9.1 Soft Skills Progress Evaluation Questionnaire

This instrument has been designed to assess the development of soft skills among Electromechanical Engineering students following the implementation of the five integral projects in the Statics course (see Table 9.1). Its primary purpose is to measure the extent to which students have strengthened competencies such as leadership, effective communication, adaptability, problem-solving, and teamwork as a result of engaging in project-based learning activities.

By systematically evaluating these skills, the questionnaire provides valuable insights into students' progress and the effectiveness of the integrative projects in promoting the acquisition of transversal competencies. The results obtained through this instrument serve both to identify areas of improvement and to guide future pedagogical strategies, ensuring that technical education is complemented by the development of interpersonal and cognitive abilities essential for successful professional performance and labor market integration.

This questionnaire will be administered at the conclusion of the intervention cycle, and its results will be comparatively analyzed against the initial diagnostic to assess both individual and group progress. By examining changes in students' responses, the study aims to identify the extent to which each soft skill—such as leadership, effective communication, adaptability, problem-solving, and teamwork—has been strengthened through the integrative projects. This comparative analysis not only provides a measure of the effectiveness of the implemented activities but also offers valuable insights for refining future pedagogical strategies, guiding targeted interventions, and supporting the continuous development of competencies that are critical for professional success and labor market integration.

Table 9.1 Soft skills progress evaluation questionnaire. Author's elaboration

	5	4	3	2	1
Section A: Effective communication					
1. I feel more confident presenting technical topics in public					
2. I write technical reports with greater clarity					
3. I improved my ability to listen and communicate within a team					
Section B: Leadership					
4. I can assume the role of technical leader with greater confidence					
5. I know how to distribute tasks and guide my team					
6. I make evidence-based decisions during project development					
Section C: Problem solving					
7. I find it easier to identify and analyze technical problems					
8. I propose practical solutions to structural challenges					
9. I can evaluate the outcomes of a solution and implement improvements					
Section D: Adaptability					
10. I adapt more easily to unexpected changes in projects					
11. I remain calm and seek solutions under new conditions					
12. I can modify my work plan without affecting the final result					
Section E: Teamwork					
13. I improved my ability to collaborate with others					
14. I recognize and value the contributions of my peers					
15. I actively help maintain group harmony and efficiency					
Section F: General opinions					
16. Which project contributed most to your personal development and why?					
17. Which skill do you consider to be the most strengthened?					
18. What suggestions would you make to improve these projects?					

Instructions: Mark the level that best represents your experience after participating in the projects. Use the scale from 1 to 5, where: **1** = Not developed at all | **2** = Slightly developed | **3** = Moderately developed | **4** = Well developed | **5** = Fully developed

9.2 Results Analysis

Table 9.2 presents the responses of ten students who participated in the integrative projects, offering a detailed illustration of their perceptions, experiences, and reflections throughout the activities. By analyzing these responses, it is possible to gain insight into how students perceived the development of their soft skills, the challenges they encountered, and the strategies they found most effective in enhancing competencies such as leadership, teamwork, problem-solving, adaptability, and communication. This qualitative information complements the quantitative data collected through the diagnostic and evaluation questionnaires, providing a richer understanding of the impact of the projects on both individual and group learning processes. Furthermore, it allows educators and researchers to identify patterns, strengths, and areas requiring further support, contributing to the continuous improvement of project-based learning initiatives within the Statics course and the broader Electromechanical Engineering curriculum.

Opinion Questions

The following section presents the students' responses to the opinion questions. These are reported verbatim to accurately reflect the participants' perspectives, experiences, and personal insights regarding the integrative projects and the development of their soft skills.

P16. Which project contributed most to your personal development and why?

1. "*The leadership project, because it allowed me to lead a team for the first time and make important decisions*".

Table 9.2 Responses to the first 15 questions of the questionnaire, using a Likert scale from 1 to 5, where 5 = fully developed and 1 = not developed at all. Author's elaboration

Student	P1	P2	P3	P4	P5	P6	P7	P8	P9	P10	P11	P12	P13	P14	P15
1	4	3	4	3	4	3	4	4	3	4	3	4	5	4	4
2	3	3	4	2	3	3	3	3	3	3	4	4	4	3	4
3	5	4	4	4	4	4	5	5	4	5	4	5	5	5	5
4	3	2	3	3	3	2	3	2	3	3	2	3	4	3	3
5	4	4	4	3	4	3	4	4	4	4	4	4	4	4	4
6	2	3	3	2	2	2	2	2	3	2	2	2	3	3	3
7	4	3	4	4	4	4	4	3	4	4	3	4	4	4	4
8	3	4	4	3	3	3	3	3	3	3	3	4	3	3	3
9	4	4	5	4	5	4	5	4	4	5	5	4	5	5	5
10	3	3	3	3	4	3	3	3	3	4	3	3	3	3	3

Questions (P): P1–P3: Effective communication; P4–P6: Leadership; P7–P9: Problem solving; P10–P12: Adaptability; P13–P15: Teamwork

Answers: 5 = Fully developed; 4 = Mostly developed; 3 = Moderately developed; 2 = Slightly developed; 1 = Not developed at all

2. *"The tower construction project, since we all worked as a team and learned to organize ourselves better".*
3. *"'Present Your Mechanism,' because it helped me speak in public, something I was very afraid of".*
4. *"'Solving Failures,' because it made me think differently in order to solve technical problems".*
5. *"I liked the adaptability project; it was a real challenge to react to the teacher's changes".*
6. *"The leadership project, since I had to organize my peers, and it was very enriching".*
7. *"The one that helped me the most was the effective communication project, now I can explain technical ideas without getting stuck".*
8. *"The collaborative tower project, because I realized how important it is to trust my team".*
9. *"I identified strongly with the problem-solving project; I love mental challenges".*
10. *"The adaptability project, because the context was changed several times, and we had to think quickly".*

P17. Which skill do you consider to be the most strengthened?

1. *Oral communication*
2. *Teamwork*
3. *Public speaking confidence*
4. *Critical thinking*
5. *Adaptability*
6. *Technical leadership*
7. *Technical writing*
8. *Cooperation*
9. *Problem solving*
10. *Decision-making under pressure.*

P18. What suggestions would you make to improve these projects?

1. *"Provide a little more time to develop the projects."*
2. *"Include examples that are closer to real work situations."*
3. *"Add a self-assessment at the end."*
4. *"Form more diverse teams in terms of skills."*
5. *"Include more support materials before starting."*
6. *"Offer personalized feedback from the instructor."*
7. *"Allow all students to rotate roles within the team."*
8. *"Provide the option to choose the type of structure to work on."*
9. *"Evaluate individual effort as well as group work."*
10. *"Incorporate examples of real industry failures."*

Overall, the statistical analysis reveals a comparative difference between the initial diagnostic results and the outcomes observed following implementation of the integral projects (see Table 9.3). The findings indicate that specific soft skills were effectively developed among the 48 second-semester students of the Electromechanical Engineering program enrolled in the Statics course.

Soft Skills Gains: Key Progress and Insights

Greatest improvement: Team collaboration (+1.06), adaptability to change (+1.04), and leadership in task distribution (+1.02) emerged as the most developed skills through the integrative projects.

Overall positive development: All questions show an increase between 0.71 and 1.06 points, indicating a significant and generalized improvement.

Sections with the highest average growth: Teamwork demonstrated outstanding progress across all subdimensions. Adaptability improved, as students learned to maintain composure and adjust plans under changing conditions. Effective communication also showed strong growth in both oral and group interaction.

Areas with lower but still significant improvement: Technical writing (Question 2) and the presentation of technical topics (Question 1) improved but to a lesser extent than other skills. This may indicate the need for greater emphasis on written communication and technical oral presentation.

Table 9.3 Results by question (ordered by greatest improvement). Author's elaboration

Question	Diagnostic Mean	Improvement Mean	Difference
13. Team collaboration	2.98	4.04	+1.06
10. Adaptation to changes	2.88	3.92	+1.04
5. Task distribution	2.98	4.00	+1.02
3. Team communication	3.17	4.12	+0.96
11. Composure under new conditions	2.88	3.81	+0.94
9. Evaluation of solutions	3.06	3.96	+0.90
14. Recognition of contributions	2.96	3.85	+0.90
7. Identification of technical problems	2.94	3.81	+0.88
6. Decision making	3.06	3.90	+0.83
4. Role of technical leader	2.98	3.79	+0.81
8. Practical solutions	2.94	3.73	+0.79
12. Modification of the work plan	2.83	3.62	+0.79
15. Group harmony and efficiency	2.81	3.60	+0.79
1. Presenting technical topics	3.10	3.85	+0.75
2. Writing technical reports	2.90	3.60	+0.71

Analysis of Section F, Qualitative Report, General Opinions

The qualitative analysis of the general opinion report focuses on examining students' subjective feedback regarding their experiences with the integrative projects. By reviewing and categorizing the participants' comments, this analysis identifies recurring themes, perceptions, and insights about the development of soft skills such as leadership, teamwork, adaptability, problem-solving, and communication. These qualitative findings complement the quantitative data, providing a richer understanding of the impact of the projects on students' learning processes and highlighting areas of strength as well as aspects that may require further attention or improvement.

Question 16: *Which project contributed the most to your personal development, and why?*

Students identified the following projects as the most significant for their personal development, as shown in Table 9.4.

Analysis: The most valued projects have a strong practical and visual component, indicating that students learn more effectively when applying theoretical concepts to real-world situations. The design of bridges and simulations foster both creativity and technical reasoning, in addition to promoting collaborative work.

Question 17: *Which skill do you consider you strengthened the most?*

In this question, it is possible to analyze what students think about the skills they believe they have developed the most (see Table 9.5).

Analysis: Effective communication was identified as the most developed skill, which aligns with the quantitative results. This suggests that the projects promoted

Table 9.4 Significant projects. Author's elaboration

Project	Frequency
Structural analysis with combined forces	12
Simulation of beams under distributed load	11
Design of a wooden stick bridge	11
Study of equilibrium in real-world structures	8
Resolution of practical cases in SolidWorks and MATLAB	6

Table 9.5 Skills strengthened through the projects. Author's elaboration

Skill	Frequency
Effective communication	16
Technical leadership	11
Problem solving	9
Teamwork	6
Adaptability	6

Table 9.6 Suggested improvements. Author's elaboration

Suggestion	Frequency
Dedicate more time to project planning	18
Include more virtual simulations	9
Include technical visits to real-world structures	9
Work with interdisciplinary projects	6
Allow greater freedom in project selection	6

increased oral and written participation, while students gained confidence in presenting technical topics. Technical leadership and problem-solving also stand out as reinforced key skills.

Question 18: *What suggestions would you make to improve these projects?*

In this section, students may present their proposals once they have completed the suggested projects, which are designed to foster the development of soft skills (see Table 9.6).

Analysis: Most students requested additional time for planning and structuring project development, reflecting their desire to achieve stronger results. Additionally, they suggested enriching projects with digital tools (simulators), field experiences, and interdisciplinary collaboration. These proposals demonstrate a broad and applied vision of their learning process.

References

1. Goleman, D.: Emotional Intelligence. Editorial Kairós (2006)
2. Robles, M.M.: Executive perceptions of the top 10 soft skills needed in today's workplace. Bus. Commun. Q. **75**(4), 453–465 (2012). https://doi.org/10.1177/1080569912460400
3. OECD.: The future of education and skills 2030 (2020). https://www.oecd.org/education/2030-project/

Chapter 10
Conclusions Part II

This chapter presents the main conclusions drawn from both the qualitative and quantitative analyses of the study, highlighting the development of soft skills among Electromechanical Engineering students through integrative projects. Based on these findings, practical recommendations are provided to guide future educational strategies, project design, and interventions aimed at further enhancing students' professional competencies and their successful integration into the labor market.

10.1 Conclusions from the Qualitative Analysis

Preference for real and tangible projects: Students valued those that allowed them to experiment and build, such as wooden stick bridges or structural analysis.

Communication as a central skill: This was the most strengthened competency, surpassing even leadership or technical problem-solving, which is particularly valuable in professional contexts where the ability to explain ideas is crucial.

Recommendations with a practical vision: Suggestions were not only feasible but also pointed toward realistic improvement, aligned with educational innovation trends such as experiential learning, interdisciplinarity, and educational technology [1, 2].

The integration of project-based educational trends responds directly to what students expressed in the open-ended questions: they desire more practical, interdisciplinary, digital projects with real-world impact. Consequently, these strategies are consistent with:

1. Project-based learning rooted in real-world applications.
2. The use of educational technology.
3. Connections with the local productive sector.
4. The strengthening of transversal competencies.

G. H. Esquivel and P. A. Hernández, *Bridging the Gap: Comprehensive Project Design for Soft Skills Development and Labor Market Integration*, Information Systems Engineering and Management 82, https://doi.org/10.1007/978-3-032-19026-0_10

These integrative projects represent a direct response to both the needs expressed by students and the competencies demanded by the industrial sector of Ciudad Juárez. Their implementation not only improves academic performance but also strengthens the graduate profile in preparation for successful and relevant workforce insertion (Ayala-Hernández & Haro-Esquivel 2024).

10.2 Recommendations

Based on the qualitative analysis of student opinions, a strong preference was identified for tangible projects in which students can experiment, build, and solve real-world problems. Moreover, effective communication emerged as the most strengthened skill following the implementation of integrative projects. This report details five strategic projects designed to foster the five soft skills most valued by employers in Ciudad Juárez when hiring electromechanical engineers: effective communication, problem-solving, technical leadership, adaptability, and teamwork. The following section presents the integrative projects designed to develop and strengthen the soft skills of students.

Building Soft Skills in IELMEC Students through Integrative Projects

Project 1: Design and Construction of a Structural Bridge

Main skill: Collaborative teamwork.

Complementary skills: Effective communication, leadership.

Students work in teams to design and construct a bridge that meets structural and material efficiency requirements. They are required to coordinate functions, allocate tasks, document their decisions, and present their technical results to a review panel.

Impact: Strengthens coordination, shared decision-making, and respect for roles within a team, simulating an industrial work environment.

Project 2: Digital Simulation of Beams under Distributed Loads

Main skill: Problem-solving.

Complementary skills: Technical communication, adaptability.

Using tools such as MATLAB, SolidWorks, or ANSYS, students model beams and analyze their behavior. They then propose structural solutions based on the results.

Impact: Stimulates analytical thinking, the ability to justify solutions, and the application of technology to address real structural challenges.

Project 3: Study of a Real Structure in the Local Environment

Main skill: Adaptability.

Complementary skills: Technical observation, problem-solving.

This project involves selecting a local structure (bridge, roof, crane), visiting the site, studying it, and proposing technical improvements or analyses. The process is documented with photographs and schematics.

Impact: Students learn to adapt to real conditions such as incomplete data or logistical restrictions, and to generate proposals based on real-world observations.

Project 4: Technical Solution to an Industry-Driven Challenge

Main skill: Technical leadership.

Complementary skills: Problem-solving, collaborative work.

In collaboration with a local company, a real case is presented (e.g., redesign of a support base). The team assumes formal roles, led by a student who manages the problem-solving process.

Impact: Promotes evidence-based leadership, well-argued decision-making, and interaction with real industrial environments.

Project 5: Public Technical Presentation before Faculty and Industry Representatives

Main skill: Effective communication.

Complementary skills: Leadership, professional confidence.

Each team presents its structural project to a mixed panel of academics and representatives of the productive sector. Evaluation focuses on clarity, technical language, and the confidence of the presenter.

Impact: Reinforces the ability to express complex ideas with precision, a key factor in interviews, technical meetings, and result presentations within the industry

References

1. Robles, M.M.: Executive perceptions of the top 10 soft skills needed in today's workplace. Bus. Commun. Q. **75**(4), 453–465 (2012). https://doi.org/10.1177/1080569912460400
2. OECD.: The future of education and skills 2030 (2020). https://www.oecd.org/education/2030-project/

Uncited References

1. Bertan Prieto, P.: The 15 types of intelligence (and their characteristics) (2023). Magazine medicoplus.com. http://medicoplus.com/psicología/tipos-inteligencia
2. Biggs, J.: Quality of University Learning. Narcea, Madrid (2006)

3. Brown, C., García, M., López, S., Rodríguez, A., Johnson, R., Smith, J.: Soft skills assessment in the workplace: strategies for effective data collection. J. Appl. Psychol. **35**(4), 432–450 (2019)
4. Caprara, G.V., Barbaranelli, C., Borgogni, L., Perugini, M.: The big five questionnaire: a new questionnaire to assess the five factor model. Personal. Individ. Differ. **15**(3), 281–288 (1993)
5. Cohen Imach, Silvina & Coronel, Claudia Paola.: Contributions of social skills theory to the understanding of violent behavior in children and adolescents. In: International Congress on Research and Professional Practice in Psychology, XVI Research Conference, Fifth Meeting of MERCOSUR Psychology Researchers. Faculty of Psychology. University of Buenos Aires, Buenos Aires (2009). https://www.aacademica.org/000-020/753
6. El Economista, Magazine.: Do you practice what you studied? El Economista. https://www.eleconomista.com.mx/el-empresario/Ejerces-lo-que-estudiaste--20120403-0165.html (2012)
7. El Heraldo de México.: 55% of professionals are unemployed or underemployed| El Heraldo de México (heraldodemexico.com.mx) (2020)
8. Escobar, A.F., Cuervo, M.R.: Design and Validation of Measurement Instruments. Pearson Education, Mexico (2008)
9. Flores, L.Y., Navarro, R.L., Lee, J.: Employment barriers and coping strategies among African American women in transition. J. Couns. Psychol. **53**(4), 441–453 (2006)
10. García Ramos, L., Páez Saavedra, J.: Engineering and Development. Colombia, Universidad del Norte. Sci. J. **30**(1), 125–142 (2012). ISSN: 0122-3461 (print). ISSN: 2145-9371
11. García, M., López, S.: Mitigating bias in soft skills assessment: strategies and best practices. J. Hum. Resour. Manag. **15**(3), 189–204 (2018)
12. García, M., Brown, C., Johnson, R., López, S., Rodríguez, A., Smith, J.: Early career experiences and their impact on workplace adaptation: a longitudinal study. J. Appl. Psychol. **38**(4) (2020)
13. Goleman, D.: Emotional Intelligence, NY (1995)
14. Goleman, D.: Emotional Intelligence. Editorial Kairós, Mexico (2001). ISBN 84-7245-371-5
15. Gómez, J.: What does emotional intelligence have to do with cars? Revista: HIGHMOTOR (2017)
16. Hazen, D.: Is engineering education in crisis? Dyna **52**(2), 47–52 (1977)
17. Hernández García, E.: The capstone project as a key factor in the labor insertion of biomedical engineering students at UPIBI. In: VIII International Congress on Educational Innovation, Mexico (2014). ISBN: 03-2014-062710460700-01. https://www.repo-ciie.dfie.ipn.mx/pdf/303.pdf
18. Indeed Mexico.: What is labor insertion and its benefits. Indeed.com Mexico (2023). https://mx.indeed.com/orientacion-profesional/desarrollo-profesional/que-es-insercion-laboral
19. Johnson, R., et al.: Developing soft skills assessment tools. J. Organ. Psychol. **25**(2), 67–79 (2018)
20. Jones, A., Brown, C.: The role of interviews in assessing soft skills. J. Workplace Psychol. **32**(4), 245–259 (2017)
21. López López, L.: View of Soft skills and their influence on the construction of meaningful learning. Ciencia Latina Multidiscip. Sci. J. (Guayaquil, Ecuador) (2021). https://ciencialatina.org/index.php/cienciala/article/view/1129/1524#:~:text=%C2%BFQU%C3%89%20SON%20LAS%20HABILIDADES%20BLANDAS,de%20manera%20efectiva%20con%20otros
22. National Employment Service.: Engineering: National Survey of Occupation and Employment. OLA (observatoriolaboral.gob.mx) (2023). https://www.observatoriolaboral.gob.mx/static/estudios-publicaciones/Ingenierias.html
23. Navarro Vera, M.: The 9 characteristics of emotional intelligence. J. Center Psychol. Speech Therapy "El teu espai" (2018). https://www.psicologosantacoloma.es/conoce-las-9-caracteristicas-de-la-inteligencia-emocional-que-todos-deberiamos-potenciar/
24. OECD.: OECD Skills Outlook 2013: first Results from the Survey of Adult Skills. OECD Publishing, Paris (2013). https://doi.org/10.1787/9789264204256-en
25. Paredes, M., Ortiz, L.: The concern for employability. Res. Train. Dev. J.: Gener. Inst. Prod. 1–15 (2019)

26. Peralta, L.P.: Soft skills questionnaire (2021). https://www.scribd.com/document/533461656/CUESTIONARIO-HABILIDADES-BLANDAS
27. Rosas, L.F., Fuentes, M.P.Q., Ramírez, J.H., López, M.T.T., Pérez, M.G.: Relationship between soft skills and the labor insertion of industrial engineering graduates from the technological Institute of Tierra Blanca. Eur. Sci. J. ESJ **18**(22), 297 (2022). https://doi.org/10.19044/esj.2022.v18n22p297
28. Sáez, C.: Performance evaluation for soft skills. SkyJobNet.com (2022). https://skyjobnet.com/evaluacion-de-desempeno-para-habilidades-blandas/
29. Sanz de Acedo, M.: Cognitive competences in higher education. In: Notebook of University Pedagogy, vol. 15, no. 29, pp. 44–53. Narcea, Madrid (2010). http://cuaderno.pucmm.edu.do/ (Dominican Republic/PUCMM/ISSN 1814-4152 (online)/ISSN 1814-4144 (print))
30. Secretariat of Public Education.: Educational Offer at the National Technological Institute of Mexico. Electromechanical Engineering. SEP, Mexico (2020)
31. Sinek, S.:. Start with why: how great leaders inspire everyone to take action (Illustrated ed.). Portfolio (2011). https://www.doonamis.es/soft-skills-que-son/#:~:text=El%20Dr.,an%C3%A1lisis%20de%20habilidades%20por%20comportamientos
32. Smith, J.: Soft skills and employment success. J. Hum. Resour. Manag. **12**(3), 45–56 (2015)
33. Thomas, J.W.: A review of research on project-based learning. The Autodesk Foundation (2000). https://www.pblworks.org/sites/default/files/litreview.pdf
34. Vega Lamilla, B., Pierina, R.: Coaching and the development of soft skills in the employees of the Company Marsh Rehder S.A., Lima (2019). https://es.scribd.com/document/533460545/HABILIDADES-BLANDAS; Soft Skills| PDF | Learning | Leadership (scribd.com)
35. Whitmore, G.: Soft skills: what they are and why they are important to find employment (donamis.es)/Empresa/August 12, 2022 (1972)
36. Zepeda-Hurtado, M.E., Cardoso-Espinosa, E.O., Rey-Benguría, C.: The development of soft skills in the training of engineers. Científica **23**(1), 61–67 (National Polytechnic Institute, Mexico) (2019)
37. Zurita, I.A.L.: Personal skills in the field of educational organizations (Soft Skills)—Indicators for measuring the development of personal skills (soft skills), 1st edn. Comillas Pontifical University (2019). https://repositorio.comillas.edu/rest/bitstreams/295564/retrieve

The manufacturer's authorised representative in the EU is Springer Nature Customer Service Centre GmbH, Europaplatz 3, 69115 Heidelberg, Germany. If you have any concerns regarding our products, please contact ProductSafety@springernature.com

Printed and bound by CPI Group (UK) Ltd, Croydon, CR0 4YY
16/09/2026
02217816-0002